Project Science

Life Processes
and Living Things

Margaret Abraitis, Angela Deighan,
Brian Gallagher, Brian Smith and Michael Toner

We hope you and your class enjoy using this book. The other books in the Project Science series are:

Materials	ISBN 978 1 897675 68 7
Physical Processes	ISBN 978 1 897675 69 4
Assessing Science Key Stage 2	ISBN 978 1 897675 35 9
Science for 5–7 Year Olds	ISBN 978 1 903853 08 5
Science and Technology for the Early Years	ISBN 978 1 903853 09 2

Published by Brilliant Publications, Unit 10, Sparrow Hall Farm, Edlesborough, Dunstable, Bedfordshire LU6 2ES, UK

Sales and stock enquiries:
Tel: 01202 712910
Fax: 0845 1309300
Email: brilliant@bebc.co.uk
Website: www.brilliantpublications.co.uk

Editorial and Marketing
Tel: 01525 222292

The name Brilliant Publications and its logo are registered trademarks.

Written by Margaret Abraitis, Angela Deighan, Brian Gallagher, Brian Smith and Michael Toner
Illustrated by Pat Murray
Cover photograph by Martyn Chillmaid

Printed in the UK

© Margaret Abraitis, Angela Deighan, Brian Gallagher, Brian Smith and Michael Toner
ISBN 978 1 897675 70 0

First published in 2000, reprinted 2007.

10 9 8 7 6 5 4 3 2

The right of Margaret Abraitis, Angela Deighan, Brian Gallagher, Brian Smith and Michael Toner to be identified as the authors of this work has been asserted by them in accordance with the Copyright, Designs and Patents Act 1988.

The following pages may be photocopied by individual teachers for class use, without permission from the publisher: 9–14, 23–42, 48–59, 63–68 and 79–96. The materials may not be reproduced in any other form or for any other purpose without the prior permission of the publisher.

Contents

Introduction — 5

Sheet no.	Title	NC link	Page no.

Life Processes — 7–14

1	Feeding	1a	9
2	Movement	1a	10
3	Growth	1a	11
4	Reproduction	1a	12
5	Living things	1a	13
6	Plants	1b	14

Humans and Other Animals — 15–42

1	Examining your teeth	2a`	23
2	Different teeth	2a	24
3	What do our teeth do?	2a	25
4	Teeth and plaque	2a	26
5	Caring for your teeth	2a	27
6	Diet and food groups	2b	28
7	Different foods	2b	29
8	A healthy diet	2b	30
9	Healthy posters	2b	31
10	The heart	2c	32
11	Circulation	2c	33
12	Pulse rate	2d	34
13	Exercise and pulse rate	2d	35
14	Skeletons	2e	36
15	Skeleton model	2e	37
16	Other skeletons	2e	38
17	Our skeletons grow	2e	39
18	Muscles	2e	40
19	Muscles and movement	2e	41
20	Working muscles	2e	42

Green Plants — 43–59

1	Plants need light	3a	48
2	Plants need water	3a	49
3	Plants need warmth	3a	50
4	Plants and leaves	3b	51
5	Roots	3c	52
6	Stems	3c	53

7	Flowering plants and seeds	3d	54
8	Dispersal of seeds	3d	55
9	Pollination	3d	56
10	Parts of the flower	3d	57
11	Light and germination	3d	59

Variation and Classification   60–68

1	Groups of living things	4a	63
2	Mini-beast key	4a	64
3	Naming mini-beasts	4a	65
4	Dog key	4a	66
5	Big cat key	4a	67
6	Prehistoric animal key	4a	68

Living Things in their Environment   69–94

1	Habitats	5b	76
2	Which habitat?	5b	77
3	Which habitat for woodlice?	5c	78
4	The wormery	5c	79
5	Animals in their habitat 1	5c	80
6	Animals in their habitat 2	5c	81
7	Adaptation	5c	82
8	What is soil?	5c	83
9	Types of soil	5c	84
10	Soil and drainage	5c	85
11	Food chains	5d	86
12	More food chains	5d	87
13	Food web	5d	88
14	Predators and prey	5d	90
15	The importance of plants	5e	91
16	Plants and animals crossword	5e	92
17	Tooth decay 1	5f	93
18	Tooth decay 2	5f	94

Glossary   95

Introduction

Welcome to what we hope is a comprehensive attempt to address the requirements of the National Curriculum (England and Wales) Key Stage 2 in a coherent way. There are three books in the series, one for each of the attainment targets:

- Life Processes and Living Things
- Materials and their Properties
- Physical Processes

Life Processes and Living Things is divided into five chapters. Each begins with background information for the teacher, summarizing and explaining key concepts which pupils are expected to acquire.

After the background information you will find a chart listing the pupil activities, the resources needed, teaching/safety notes for each of the activities, and answers for the pupil pages. The answers are the writers' own results. In the complicated, but exciting world of science, these may be different from your results, but that does not make you wrong! Sometimes there will be small variations between the way in which you and the writers carried out the activities. This does not matter, and can make a good discussion point with pupils. The question 'why?' is a key part of science.

Pupils will need a pencil or pen for all the activities and additional sheets of paper are necessary for some. Other resources required are listed under 'Resources needed', and again at the top of each pupil page.

The 'Teaching/safety notes' give additional information and highlight areas where particular care is required. Before attempting any of the activities suggested, you should be fully conversant with the Association for Science Education's *Be Safe!* booklet and any guidelines laid down by your local authority/education department/board of governors, etc.

The bulk of the book consists of pupil sheets that may be photocopied free of charge by the purchasing institution. The sheets provide a series of exciting investigations to reinforce the concepts in the pupils' minds. The activities closely adhere to QCA's Scheme of Work for Science and provide a practical, useful way for the teacher to administer the Scheme of Work.

To make the pupil sheets easy to use we have used logos along the left margin to flag the different types of text:

 Passages the pupil should read in order to find out more about the topic

 Investigations, experiments, puzzles, and other activities

 The pupil has to write an answer down

 Activities which stretch the pupil. Many require the use of reference books and/or CD-ROMs.

The contents page indicates which section of the National Curriculum Programme of Study is addressed on each pupil sheet.

Where practicable, you should attempt the activities yourself before introducing them to your pupils. You should also ensure that the equipment you intend to use is suitable for the activity/activities and the children using it.

We must emphasize that all activities should be carried out under teacher supervision. The publishers do not accept liability for any injury or damage howsoever caused arising from activities suggested in this publication.

No scheme can give teachers all the answers. However, we hope that this scheme will enable you to use your time where it can be spent most productively – namely helping individual pupils rather than doing repetitive planning and preparing worksheets.

We hope you and your pupils will enjoy **Project Science**.

To obtain a copy of *Be Safe!* contact the Association for Science Education, College Lane, Hatfield, Hertfordshire AL10 9AA.

Life Processes

Background information

There are various ways to define what constitutes a living **organism** (the scientific name of something that is alive). It is, however, useful to think of living things in terms of what they have in common. Living things such as animals and plants grow, reproduce, feed, move, take in and pass out gases and react to their environment.

It is easy to think of animals moving, breathing and reacting to their surroundings, but plants also share these characteristics. Plants cannot move from place to place, but some open flowers during the day and close them at night. Some turn to face the Sun. All green plants take in carbon dioxide and give out oxygen during **photosynthesis**. Many plants react to changes in sunlight which causes them to make seeds and reproduce.

Photosynthesis is the process by which a plant can change light energy from the Sun into chemical energy (food). A chemical called **chlorophyll** is required for this reaction. Chlorophyll gives plants their green colour.

Helpful words	
Organism	Chlorophyll
Photosynthesis	Carbon dioxide
Reproduce	Oxygen
Glucose	

Photosynthesis

$$\text{Water} + \text{Carbon dioxide} \xrightarrow[\text{Chlorophyll}]{\text{Light energy}} \text{Oxygen} + \text{Glucose}$$

The glucose produced by photosynthesis can be changed to starch, protein and fats. Since all animals, including humans, are dependent on plants for food, this reaction is fundamental for the survival of all animal life.

Some ideas for further work

◆ Some animals have life cycles which involve them going through a metamorphosis. Find out what this is and the names of one or more animals which go through this process.

Life Processes

Activity	Answers	Resources needed	Teaching/safety notes
Sheets 1–5 (pages 9–13) Feeding, Movement, Growth, Reproduction and Living things	• Things which feed, move, grow, reproduce and are living things: horse, snail, boy, dog, cow and owl. • Things which do not feed, move, grow, reproduce and which are non-living things: pencil, bed, book, television, hammer and shoe.	Scissors Glue	
Sheet 6 (page 14) Plants	The plants which are given fertilizer should grow faster. Plants require nutrients such as nitrogen and phosphorous for healthy growth. Lack of these in the soil will result in poor plant growth. Fertilizers can be used to add these nutrients to the soil.	2 trays of soil Fertilizer Young plants	Fertilizers must be handled with care as they could be poisonous, flammable, corrosive or irritant. Household fertilizers such as Baby Bio can be used under close adult supervision.

Feeding

Life Processes Sheet 1

You need: scissors and glue.

 **Activity** Cut out and sort the things below into things which feed and things which do not feed.

Things which feed	Things which do not feed

Movement

Life Processes Sheet 2

You need: scissors and glue.

Activity: Cut out and sort the things below into things which can move on their own and things which cannot move on their own.

Things which can move on their own	Things which cannot move on their own

Life Processes and Living Things

© M Abraitis, A Deighan, B Gallagher, B Smith & M Toner
This page may be photocopied for use by the purchasing institution only.

Life Processes
Sheet 3

Growth

You need: scissors and glue.

Activity Cut out and sort the things below into things which grow and things which do not grow.

Things which grow	Things which do not grow

© M Abraitis, A Deighan, B Gallagher, B Smith & M Toner
This page may be photocopied for use by the purchasing institution only.

Life Processes and Living Things
11

Reproduction

Life Processes
Sheet 4

You need: scissors and glue.

 **Activity** Cut out and sort the things below into things which can produce young and things which cannot produce young.

Things which can produce young	Things which cannot produce young

Life Processes and Living Things

Living things

Life Processes Sheet 5

You need: scissors and glue.

 **Activity** Living animals feed, move, grow and produce young. Cut out and sort the things below into living things and non-living things.

Living things	Non-living things

© M Abraitis, A Deighan, B Gallagher, B Smith & M Toner
This page may be photocopied for use by the purchasing institution only.

Life Processes
Sheet 6

Plants

You need: 2 trays of soil, fertilizer and some young plants.

 **Read** Plants are living things and can do many of the things that animals do. Unlike animals, however, plants make their own food.

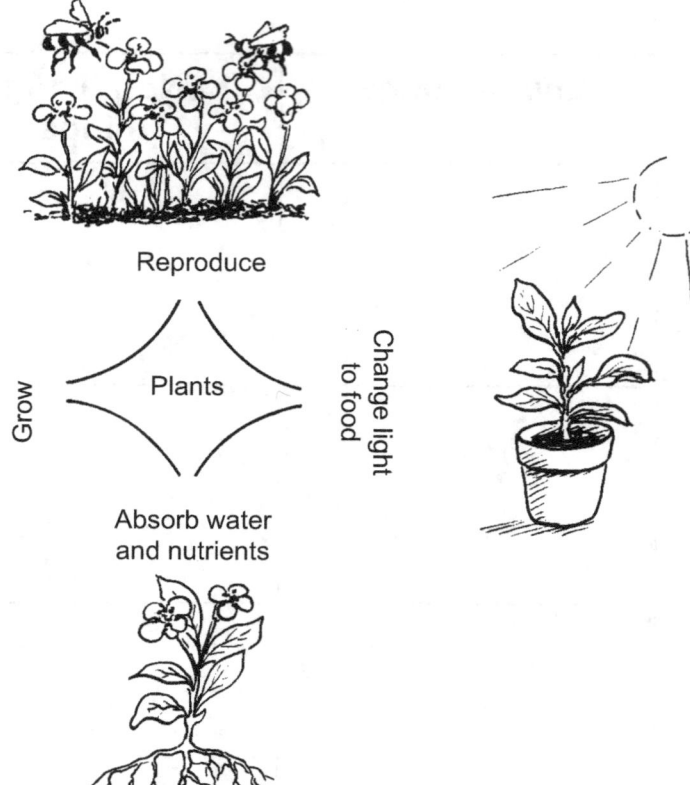

 **Activity** Investigate whether fertilizers affect the growth of young plants. Grow some young plants in a tray containing normal soil and some in a tray containing soil mixed with fertilizer.

Measure growth each week	At the start	After one week	After two weeks	After three weeks
with fertilizer				
without fertilizer				

Life Processes and Living Things

© M Abraitis, A Deighan, B Gallagher, B Smith & M Toner
This page may be photocopied for use by the purchasing institution only.

Humans and Other Animals

Background information

Teeth

A grown adult should have 32 teeth, 16 on the top jaw and 16 on the bottom jaw. There are three types of teeth. The front ones are called **incisors**, which are used for cutting food. **Canines** are sharp and pointed for biting down on food. **Molars** (the front smaller molars are called **pre-molars**) are used for chewing. There should be the same number and type of each tooth on both the bottom and top jaws.

Animals like dogs have large and very sharp canine teeth for biting, gripping and tearing meat. (*Canine* is from the Latin word for dog.) Other animals, such as sheep, have sharp incisors for biting plants and large molars for grinding the plants down. The bottom jaw of a sheep moves from side to side, which helps it to grind down its food more easily.

Teeth are living parts of our bodies so they require a blood supply and contain nerves to give sensation. The white outer part of a tooth is called enamel. The enamel surrounds and protects the living part of the tooth, which is called dentine. In the middle of the tooth is a cavity called the pulp cavity which contains blood vessels and the nerves. Acid produced by bacteria feeding on sugar left on the teeth after eating can soften the enamel and sometimes form a hole in it. If this hole goes right through to the living part of the tooth, we feel pain due to air reaching the nerve. The tooth can be protected by brushing food particles away and by strengthening the enamel by drinking water which contains small amounts of fluoride.

Helpful words

Teeth
- Incisor
- Canine
- Molar
- Enamel
- Dentine
- Nerve
- Pulp
- Fluoride

Nutrition
- Carbohydrate
- Protein
- Fat
- Vitamins
- Mineral
- Diet
- Starch

Circulation and movement
- Artery
- Vein
- Muscle
- Pulse
- Heartbeat
- Skeleton
- Vertebrates
- Invertebrates
- Joint
- Tendon

Nutrition

Foods can be grouped into certain classes depending on their function in the body, as shown in the table below.

Food class	Function in the body
Carbohydrates (starches + sugars)	These provide energy for activity.
Protein	This food is essential for growth and repair.
Fats	They are for reserve energy and are stored around the body.
Vitamins + minerals	These are needed to prevent disease and for some chemical reactions.

Life Processes and Living Things

Humans and Other Animals

The food we eat contains varying amounts of each of the food classes on page 15 and it is important that we eat a varied and balanced amount of them if we are to stay healthy and fit. The table below shows some of the foods which contain mainly these food classes.

Food class	Food examples
Starches	Bread, potatoes, rice, pasta and cereals
Sugars	Jams, sweets, cakes and fizzy drinks
Protein	Fish, meat, cheese, eggs and milk. Soya beans and peas
Fats	Chips, crisps, cream and butter
Vitamins + minerals	Bread, fish, milk and fruit

Many foods, such as vegetables, fruit and cereals, contain fibre which is not digested. Fibre (roughage) performs an important role in aiding the passage of food through the digestive system.

Circulation

The **heart** is divided into four main chambers and has a wall made of thick muscle.

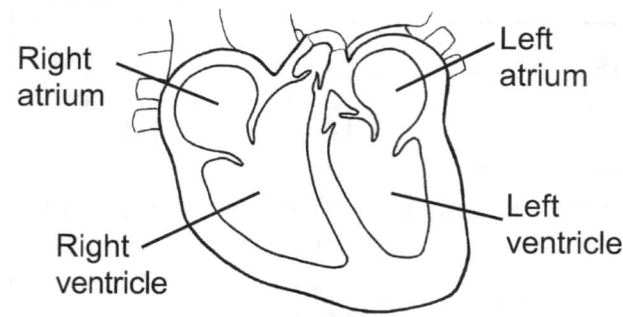

The function of the heart is to pump blood, which contains many substances dissolved in it, including oxygen, around the body. As the heart contracts, it squeezes blood out; as it relaxes, it fills up with blood again. The blood is pumped by the right-hand side of the heart to the lungs where the oxygen is absorbed. The oxygen-rich blood then passes from the lungs to the left-hand side of the heart. It is then pumped around the body, the oxygen being absorbed as it circulates. The blood then returns to the right-hand side of the heart and the process is repeated.

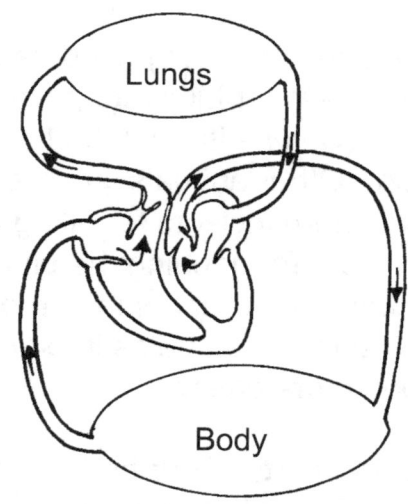

Arteries carry blood away from the heart to the rest of the body and **veins** carry blood back from the body to the heart. Arteries divide into smaller and smaller capillaries which deliver the blood to the body's living cells. The blood itself is mostly made from a watery liquid which is called **plasma**. The plasma contains many dissolved chemicals which have their own particular uses in the body. Blood also contains red blood cells, which carry oxygen to the body cells, white blood cells, which kill germs, and platelets, which form blood clots and reduce blood loss if the body is cut.

Humans and Other Animals

When the heart beats, it forces blood through the arteries which swell slightly at each heartbeat. If you put your finger on your wrist you can feel your pulse. The pulse is caused by the blood flowing through the artery each time the heart beats. You can therefore count how often your heart beats per minute by counting the number of times you feel your pulse each minute. A typical pulse rate for someone at rest would be about 72 beats per minute. When we exercise our muscles require more energy, which they get from food and oxygen in the blood. To supply more blood our heart beats more often and our pulse rate rises. After exercise the pulse rate returns to normal. The fitter a person is, the quicker their pulse rate will return to normal (recovery time). Also, a fit person's heart can supply larger quantities of blood, so their pulse rate will not increase as much during exercise as an unfit person's.

Movement

There is an average of 206 bones in an adult human body. Babies are born with over 300 bones, some of which fuse together as they get older. The skeleton has three main functions: to **support** the body so that it can stand; to **protect** vital organs such as the heart, brain and lungs; and, along with the muscles, to allow **movement**. The smallest bone in the body is the stapes (stirrup) bone, which is one of three bones found in the ear. Animals such as humans, which have backbones are called **vertebrates** and animals which do not have backbones are called **invertebrates**. Joints are where the bones meet and allow the limbs to bend. There are different types of joints. For example, the joint at the elbow is called a hinge joint and that at the hip is called a ball and socket joint.

There are between 656 and 850 (depending on how you count them) muscles in the human body. Voluntary muscles (which are connected to the skeleton) allow us to move at will, and involuntary muscles (like the ones in our intestines) perform their function on their own without our control. Muscles are connected to bones by tough fibres called tendons. Signals from the brain cause muscles to contract, pulling the limb and causing movement. In the arm, the biceps muscle contracts and lifts the arm and the triceps muscle contracts and lowers the arm. The biceps and triceps muscles act as a pair to produce movement.

Some ideas for further work

◆ Look at pictures of models of the teeth of different animals such as cats and sheep. Discuss why their teeth are different from our own.

◆ Devise you own healthy diet for a month. Write a chart for all meals and snacks for each day.

   1. Keep a record to show how faithfully you keep to your healthy diet.
   2. Write a paragraph on how you feel it has benefitted you.
   3. Write another paragraph on how this experiment might benefit you in the future. (For example, has it made you more aware of the importance of a proper diet?)

Humans and Other Animals

◆ What sort of things can lead to heart disease and how can the chances of heart disease be reduced?

◆ Compare the skeletons of other animals with the human skeleton. Some bones, such as in a bird, are lighter than similar-sized bones in animals that do not fly. Why is this?

◆ Find out the names of some diseases which are caused by the lack of a good balanced diet. What happens if we do not eat enough or eat too much?

Humans and Other Animals

Activity	Answers	Resources needed	Teaching/safety notes
Sheet 1 (page 23) Examining your teeth	Children's first teeth start to grow when they are about six months old. Four-year-olds should have 20 milk teeth. A couple of years later four extra teeth will grow. As children get older, their permanent adult teeth begin to grow and push the milk teeth out. • Top teeth showing in illustration from left to right: incisor, incisor, incisor, canine, pre-molar, pre-molar, molar. • Bottom teeth showing in diagram from left to right: incisor, incisor, incisor, incisor, canine, pre-molar, pre-molar, molar, molar.	Plastic mirror	The children could use their tongues to count their teeth.
Sheet 2 (page 24) Different teeth	• Top teeth: molars – 6; pre-molars – 4; canines – 2; incisors – 4. • Bottom teeth: the same. There are 32 teeth in a full set of adult teeth.	Coloured pencils	There should be the same number of each type of tooth on the upper and lower jaw. The third molars are the wisdom teeth. Wisdom teeth are no longer useful as our jaws have changed shape as we have evolved, leaving no room for these molars.
Sheet 3 (page 25) What do our teeth do?	See Sheet 2 for the number of teeth an adult should have. • Incisor: biting, cutting. • Canine: biting, tearing. • Pre-molar: chewing, grinding. • Molar: crushing, grinding.	Apple Bread Crisps	
Sheet 4 (page 26) Teeth and plaque		Disclosing tablet Toothbrush Water Mirror Coloured pencils	The disclosing tablet will stain the plaque. Brushing will remove the plaque and the stain along with it. The children should bring in their own toothbrushes from home. They should not share toothbrushes.

Life Processes and Living Things

Humans and Other Animals

Activity	Answers	Resources needed	Teaching/ safety notes
Sheet 5 (page 27) Caring for your teeth	Good things for the care of teeth: brushing with toothpaste, using mouthwash, eating fruit and vegetables, visiting the dentist. Bad things: eating cakes, chocolate and sweets.		
Sheet 6 (page 28) Diet and food groups	If we did not eat we would eventually die of starvation.		Our diets should contain the right amount of carbohydrates, fats, proteins, vitamins, minerals and fibre. If we do not eat a proper balance of these foods, then we will suffer from malnutrition.
Sheet 7 (page 29) Different foods	• Growth foods: chicken, steak, egg, fish. • Activity foods: bread, pasta, sweets, rice.	Scissors Glue	**Growth foods** are those foods containing a lot of protein and **activity foods** are carbohydrates (sugar and starch) and fats. Most foods, like a fatty steak, contain more than one food group.
Sheet 8 (page 30) A healthy diet	The children might include the following in their table. • Breakfast: toast and butter – starch and fats, activity foods. • Lunch: burger, chips, fizzy drink – meats, starch, fats and sugar, growth and activity foods. • Dinner: chicken, potatoes, carrots and an apple – meat, starch, vegetables and fruit, growth, activity and fibre foods. • Snacks: crisps and an orange – fats and fruit, activity and fibre foods.		
Sheet 9 (page 31) Healthy posters		Art materials Large sheets of paper	

Humans and Other Animals

Activity	Answers	Resources needed	Teaching/safety notes
Sheet 10 (page 32) The heart	• Top left: right atrium • Top right: left atrium • Bottom left: right ventricle • Bottom right: left ventricle.		The heart circulates around 9000 litres of blood a day.
Sheet 11 (page 33) Circulation		Coloured pencils/pens	
Sheet 12 (page 34) Pulse rate	Possible values could be: • Attempt 1 – 68 beats per minute. • Attempt 2 – 76 beats per minute. • Attempt 3 – 72 beats per minute.	Stop clock/timer	
Sheet 13 (page 35) Exercise and pulse rate	Doing exercise causes an increase in our pulse rate. Possible values could be (from left to right across table): 72, 110, 105, 98, 86.	Stop clock/timer	
Sheet 14 (page 36) Skeletons	• Left from top: Ribs, Pelvis, Thigh bone (femur), Shin bones (tibia and fibula). • Right from top: Skull, Upper arm bone (humerus), Backbone (spinal column), Lower arm bones (radius and ulna).	Reference materials	
Sheet 15 (page 37) Skeleton model		Scissors Glue Card Paper fasteners	Depending on how they are counted, the average person has 206 bones. If possible, the pupil sheets should be enlarged on the photocopier to make the model easier to build.

Life Processes and Living Things

Humans and Other Animals

Activity	Answers	Resources needed	Teaching/safety notes
Sheet 16 (page 38) Other skeletons	The five skeletons all have skulls, backbones, ribs and limbs.		The bones of a vertebrate are needed to support the body, protect vital organs and to give an anchor for tendons connected to muscles. Animals such as birds have very light bones; other animals have thick heavy bones.
Sheet 17 (page 39) Our skeletons grow		Reference materials	The average boy (US statistics) is taller than the average girl. However, at about the age of 12, girls outgrow boys. By the age of about 14, the growth rate of boys increases. On average, boys grow taller than girls.
Sheet 18 (page 40) Muscles	• Left, from top: Biceps, Quadriceps. • Right, from top: Deltoid, Pectoral, Tibialis.	Reference materials	The **biceps** muscle bends the arm at the elbow. (There are also biceps at the back of the thigh.) The **deltoid** muscle raises the arm sideways and also twists the arm. The **quadriceps** muscle has four parts and straightens the knee. The **tibialis** muscles flex the ankle and raise the toes. The **pectoral** muscles are in the chest and allow certain arm and shoulder movements.
Sheet 19 (page 41) Muscles and movement	The children should be able to feel their muscles contract (flexing) and relaxing (extending).		A muscle like the biceps is able to contract. When this happens, the arm is raised at the elbow and the triceps is lengthened. The biceps is unable to lengthen on its own so, to lower the arm, the triceps contracts. The biceps is pulled to its orginal length and the arm lowers. Muscles working together like this are called an **antagonistic** pair.
Sheet 20 (page 42) Working muscles	As above.	Weight	

Humans and Other Animals Sheet 1 — Examining your teeth

You need: a plastic mirror.

 **Read**

Our first set of teeth is only a temporary set. They are called milk teeth. Normally there are 20 milk teeth and they last only a few years. A new adult set grows when they fall out.

 **Activity**

- Examine your teeth using a mirror.
- Look at the different shapes and sizes of each tooth.
- Count how many teeth you have.

 **Answer**

Write about what you found. Do you have any milk teeth left?

 **Activity**

Match the three other types of teeth with their position in the mouth.

Name		Shape
Incisors		
Canines		
Pre-molars		
Molars		

© M Abraitis, A Deighan, B Gallagher, B Smith & M Toner

Different teeth

Humans and Other Animals
Sheet 2

You need: coloured pencils.

 Look at the full set of adult teeth below.

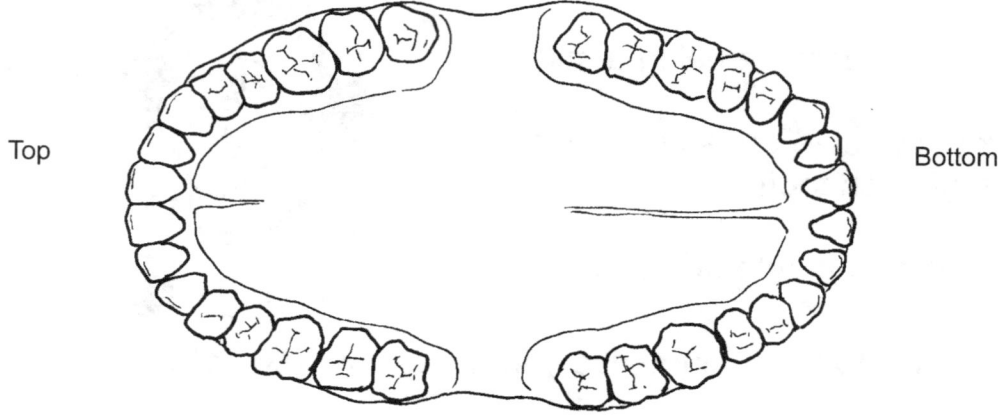

Top   Bottom

◆ Colour the   incisors **red**
                canines **blue**
                pre-molars **green**
                molars **yellow**

◆ Fill in the following table which records the number of each type of tooth in a full set.

Type of tooth	Number of top teeth	Number of bottom teeth
Incisor		
Canine		
Pre-molar		
Molar		

 How many teeth should there be in a full set of adult teeth?

Life Processes and Living Things

What do our teeth do?

Humans and Other Animals
Sheet 3

You need: an apple, a piece of bread and some crisps.

Examine your own teeth again and fill in the table to record the number of each type of tooth **you** have.

Type of tooth	Number of top teeth	Number of bottom teeth

Look again at Sheet 2, Different teeth. Do you have the same numbers as in the table?

If not, why not?

Let's find out what our different teeth do.

◆ Eat each food slowly.

◆ Think about what each type of tooth is doing as you are eating (which type is biting, which type is crushing food, chewing, etc).

◆ Use the words biting, chewing, crushing, tearing, grinding and cutting.

Type of tooth	What is it used for when eating?
Incisor	
Canine	
Pre-molar	
Molar	

Humans and Other Animals
Sheet 4 — Teeth and plaque

You need: one disclosing tablet, a toothbrush, water, a mirror and coloured pencils.

Read

Our teeth can become bad if we forget to look after them properly. If we forget to brush our teeth, a substance called **plaque** builds up on them. **Plaque** attacks our teeth and causes **tooth decay**.

Activity

Chew the disclosing tablet and then look at your teeth in the mirror.

What do you notice about your teeth?

The pink stains show areas where plaque has built up. Now brush your teeth – especially in the pink areas.

Look further

Show what would happen to your teeth if you never brushed them. Colour in the picture of the tooth with coloured pencils.

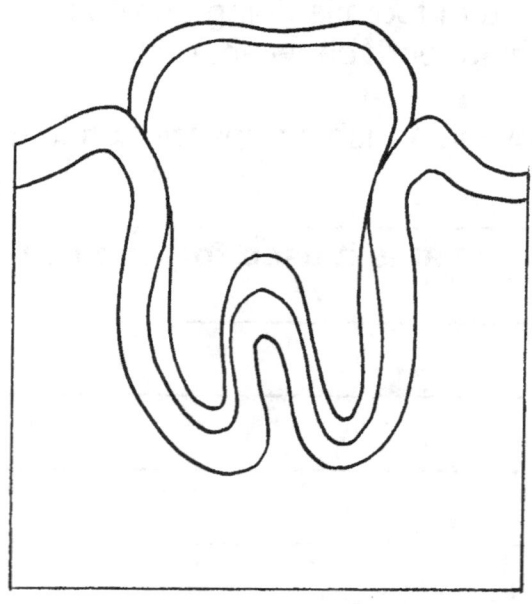

Life Processes and Living Things

Humans and Other Animals
Sheet 5

Caring for your teeth

Activity — Use the pictures below to help you write about the different ways to take care of your teeth.

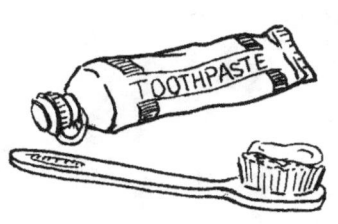
Toothpaste

Mouthwash

Cakes

Fruit

Chocolate

Vegetables

Brushing

Sweets

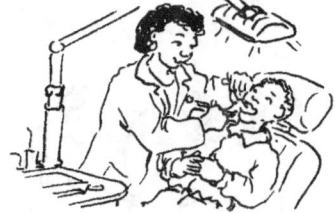
Visits to the dentist

Answer

© M Abraitis, A Deighan, B Gallagher, B Smith & M Toner

Humans and Other Animals
Sheet 6 — Diet and food groups

Read — **Diet** is a description of the food we eat regularly. All animals, including humans, need to eat a sensible, balanced diet.

Answer — What would happen if we didn't eat?

Foods are sorted into the food groups shown below.

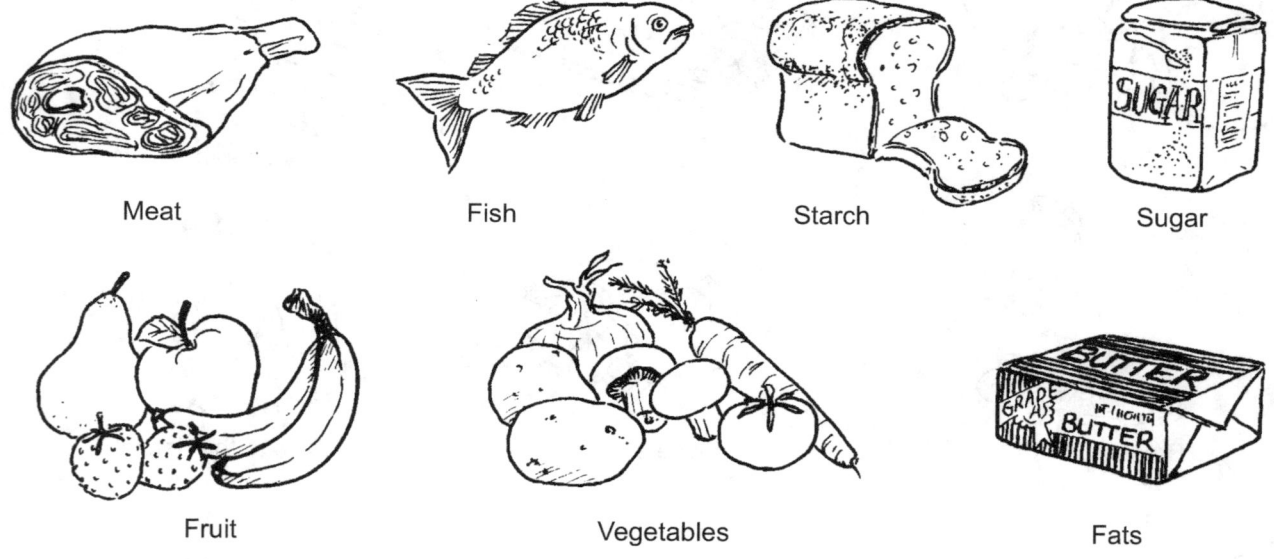

Meat   Fish   Starch   Sugar

Fruit   Vegetables   Fats

In supermarkets, food is usually stacked in groups. Apart from making it easier to find things, this is because different food groups do different jobs to keep us alive and healthy.

Look further — Design a supermarket shelf to display examples of the types of food shown above. Use a separate sheet of paper.

Life Processes and Living Things

Different foods

Humans and Other Animals — Sheet 7

You need: scissors and glue.

 **Read** — Foods can be grouped according to their uses.

Activity foods	Fibre foods	Healthy growth foods
Starch	Vegetables	Meat
Sugar	Fruit	Fish
Fats	Your example	Your example

 **Activity** — Cut out and then sort the foods below into **growth foods** and **activity foods**.

Chicken, Bread, Steak, Egg, Pasta, Sweets, Fish, Rice

A healthy diet

Humans and Other Animals
Sheet 8

 It is very important to eat a healthy balanced diet. This means you should eat foods from all food groups and have enough growth foods and activity foods.

Growth foods

Activity foods

 List all the foods you ate yesterday in the table.

Meat	Food(s)	Food group(s) (meat, fish, starch, sugar, fruit, fat or vegetables)	Is it an *activity* food, a *growth* food or *other* type of food?
Breakfast			
Lunch			
Dinner			
Snacks			

Life Processes and Living Things

Healthy posters

Humans and Other Animals — Sheet 9

You need: art materials and large sheets of paper.

 Design two healthy posters for your doctor's and dentist's surgeries.

Healthy eating – a fact poster to explain the importance of eating a healthy, balanced diet.

Healthy teeth – a fact poster to explain the importance of teeth and how to look after them.

Plan your ideas here.

Healthy eating	Healthy teeth

© M Abraitis, A Deighan, B Gallagher, B Smith & M Toner
This page may be photocopied for use by the purchasing institution only.

Life Processes and Living Things

Humans and Other Animals
Sheet 10
The heart

Read

The heart is a hollow muscle which pumps blood around the body through tubes called **blood vessels**. The large blood vessels which carry blood away from the heart are called **arteries**. Those carrying the blood back again are called **veins**.

Activity

Use the words below to label the diagram of the heart.

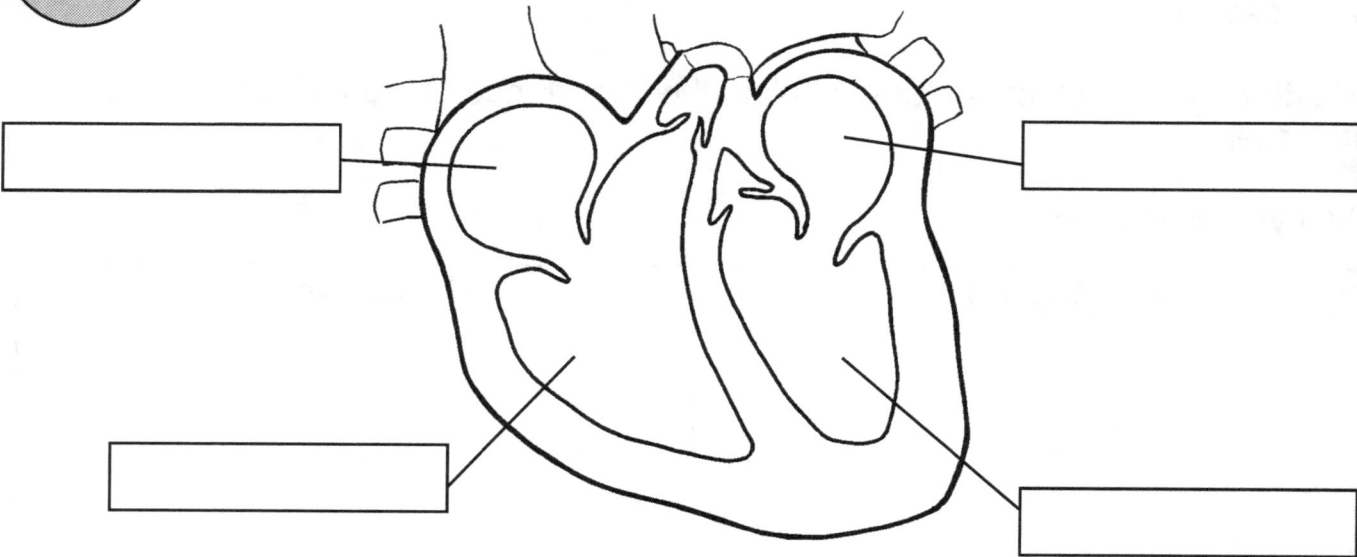

| Right ventricle | Left ventricle | Left atrium | Right atrium |

Look further

Can you find the heart in the picture below? Colour it in.

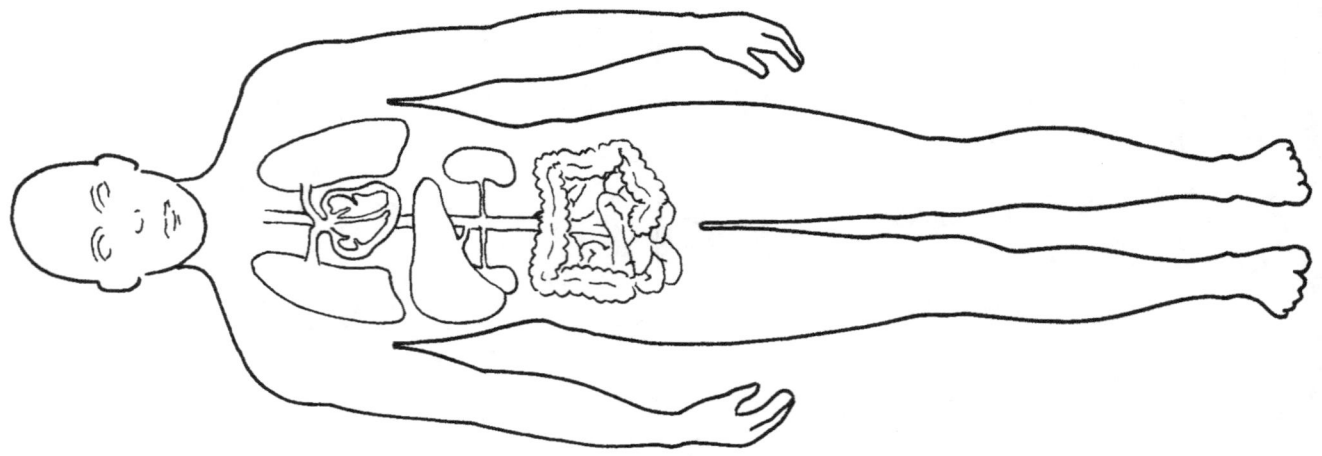

Life Processes and Living Things

Humans and Other Animals
Sheet 11

Circulation

You need: coloured pens or pencils.

Activity — Use different colours to trace the blood flow from the heart and through the arteries to the lungs, liver, kidneys, feet and the brain.

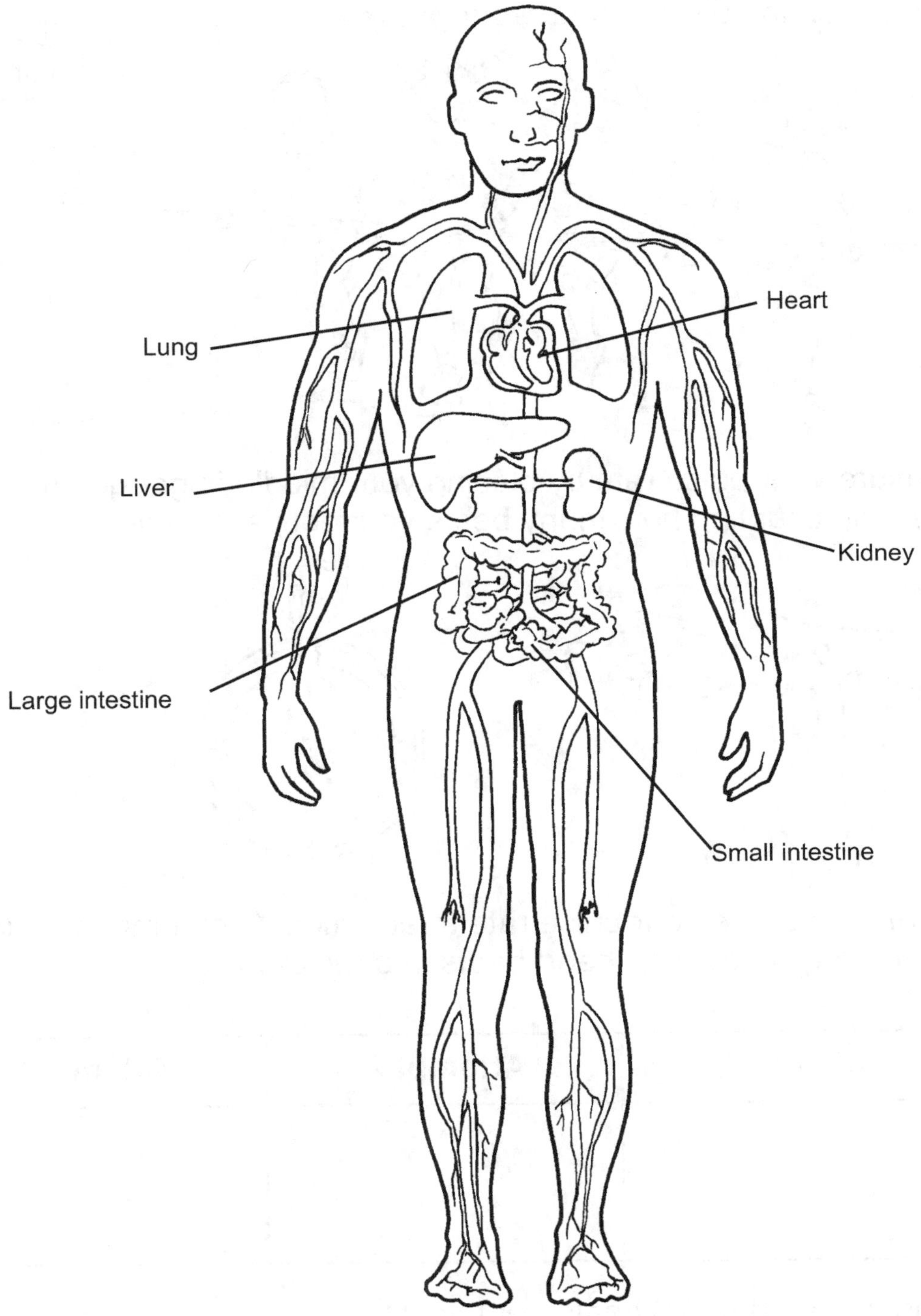

- Lung
- Heart
- Liver
- Kidney
- Large intestine
- Small intestine

© M Abraitis, A Deighan, B Gallagher, B Smith & M Toner
This page may be photocopied for use by the purchasing institution only.

Pulse rate

Humans and Other Animals
Sheet 12

You need: a stop clock or timer.

 **Read** Blood is constantly pumped round our body by our heart. When we exercise, our muscles need more blood for energy and oxygen.

To supply more blood to our muscles, our heart beats faster. We can measure how fast our heart is beating by measuring our **pulse rate**.

Activity ◆ Measure your pulse rate by placing your middle finger gently on your wrist and counting how many beats you feel per minute.

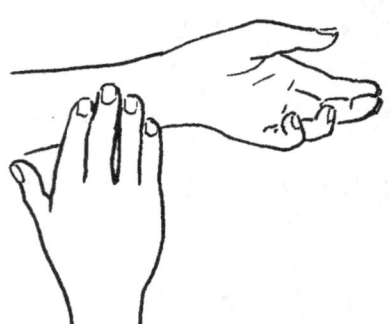 

◆ Sit down, relax and measure your pulse rate three times. (Remember, your pulse rate is how many times your heart beats in a minute.)

	Attempt 1	Attempt 2	Attempt 3
Pulse rate in beats per minute			

Your pulse rate should be about 70 beats per minute.

Life Processes and Living Things

Exercise and pulse rate

Humans and Other Animals — Sheet 13

You need: a stop clock or timer.

**Activity**   Let's investigate how doing exercise affects our pulse rate.

◆ Measure your pulse rate when you are sitting down and relaxed.

◆ Do some sit-ups for 1 minute and then measure your pulse rate again.

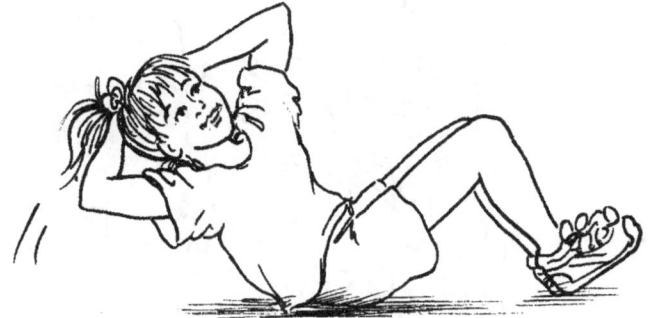

Pulse rate ____ beats per minute   Pulse rate ____ beats per minute

**Answer**   How does doing exercise affect your pulse rate?

**Look further**   Everyone's pulse rate increases when they do exercise. The fitter the person, then the quicker their pulse rate returns to normal after exercise.

Investigate how long it takes for your pulse rate to return to normal after doing 1 minute of exercise.

Pulse rate before exercise	Pulse rate immediately after exercise	Pulse rate 1 minute after exercise	Pulse rate 2 minutes after exercise	Pulse rate 3 minutes after exercise

© M Abraitis, A Deighan, B Gallagher, B Smith & M Toner
This page may be photocopied for use by the purchasing institution only.

Life Processes and Living Things

Humans and Other Animals
Sheet 14

Skeletons

 Label the bones in the skeleton with the correct words below.

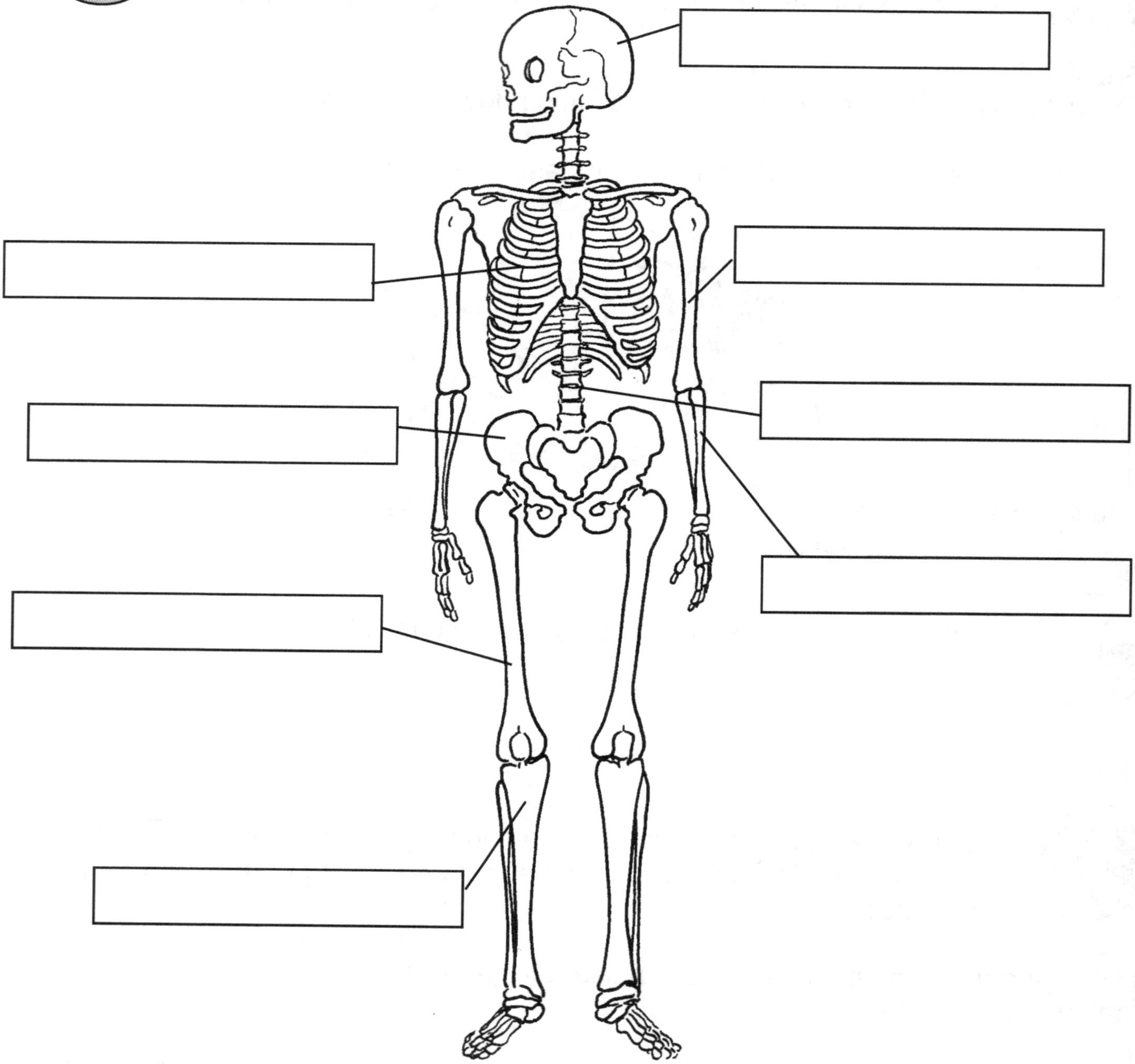

Upper arm bone	Lower arm bones	Pelvis
Backbone	Ribs	Thigh bone
Skull	Shin bones	

Life Processes and Living Things

Skeleton model

Humans and Other Animals
Sheet 15

You need: scissors, glue, card and some paper fasteners.

Activity

Build your own skeleton using the cut-out model below. Stick to thick card, cut along the line and then cut out each part.

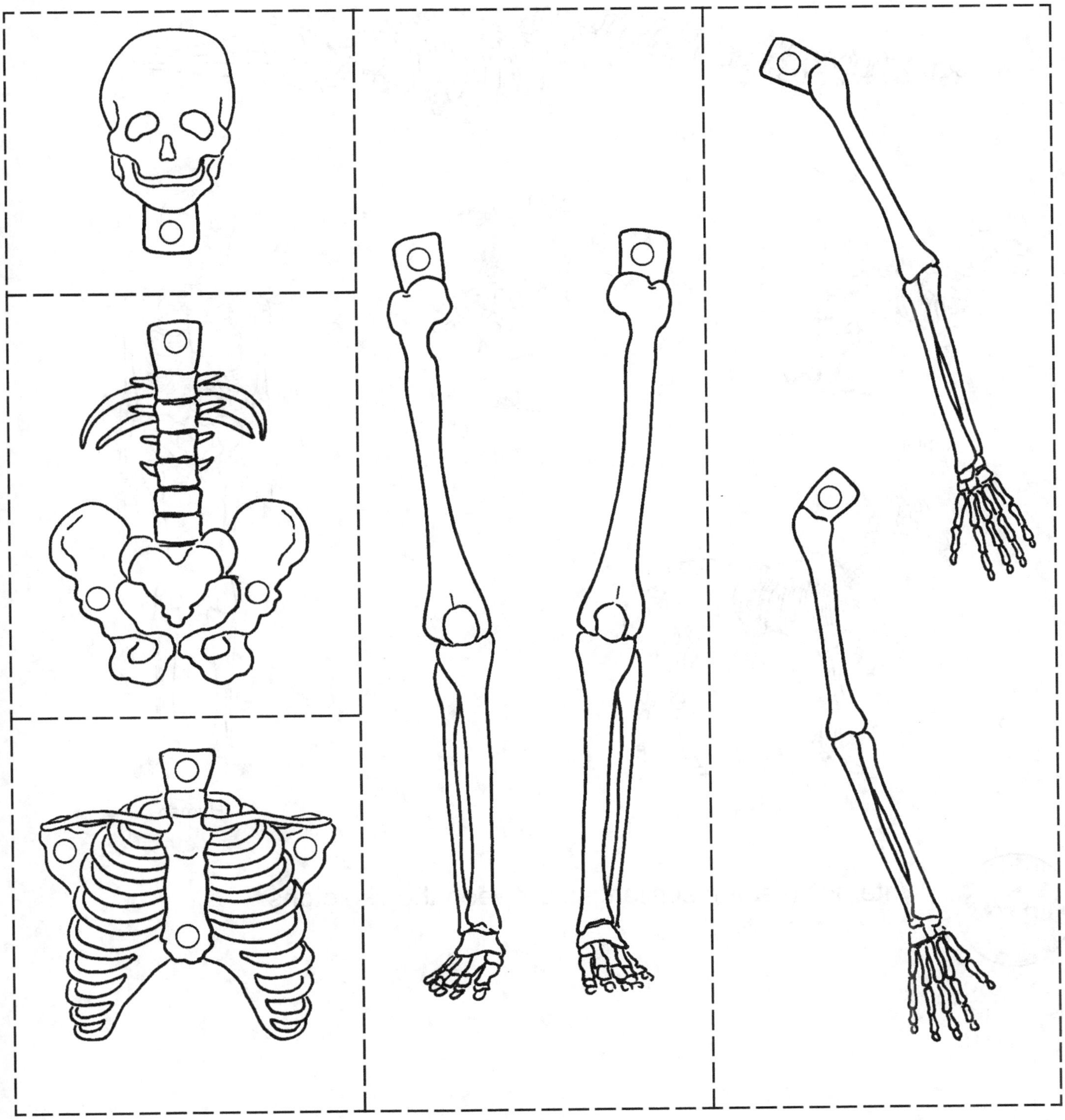

© M Abraitis, A Deighan, B Gallagher, B Smith & M Toner
This page may be photocopied for use by the purchasing institution only.

Life Processes and Living Things

Other skeletons

Humans and Other Animals
Sheet 16

Humans are not the only animals with skeletons.

Compare the skeletons of the other animals drawn below. (They are not drawn to scale).

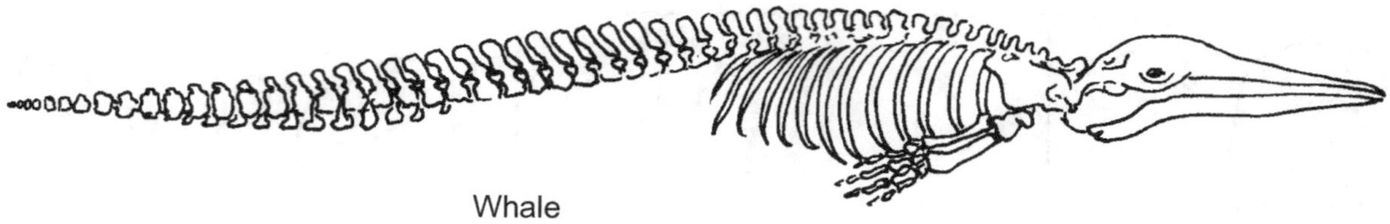

Whale

Rabbit

Bird

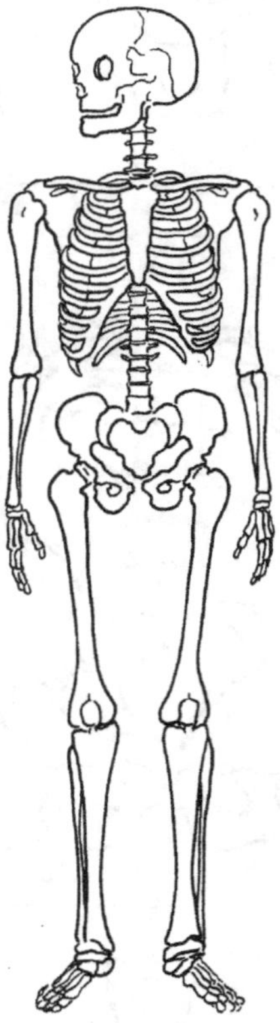

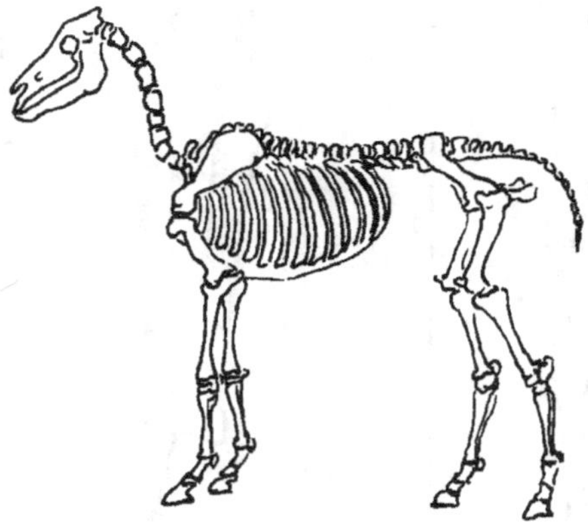

Horse

Human

Write down some similarities between the skeletons.

Our skeletons grow

Humans and Other Animals
Sheet 17

 **Read** Shortly after a baby is born, a doctor or midwife will measure the baby's head circumference, body length and weight.

These measurements can give an idea of how well the baby is developed at birth, and will be used to check on the baby's future growth.

Activity The children below have their own ideas about how bones grow. Investigate one or all of their ideas.

> Girls have longer arms than boys.

> Our bones stop growing when we are 10.

> Bones are light and soft.

© M Abraitis, A Deighan, B Gallagher, B Smith & M Toner
This page may be photocopied for use by the purchasing institution only.

Life Processes and Living Things

Muscles

Humans and Other Animals
Sheet 18

 Our bodies have lots of **muscles** which pull joints to move different parts of our body.

Look at Harry the Muscle-Man below. Nearly every muscle in his body can be seen.

 Find out the names of the muscles the arrows are pointing to.

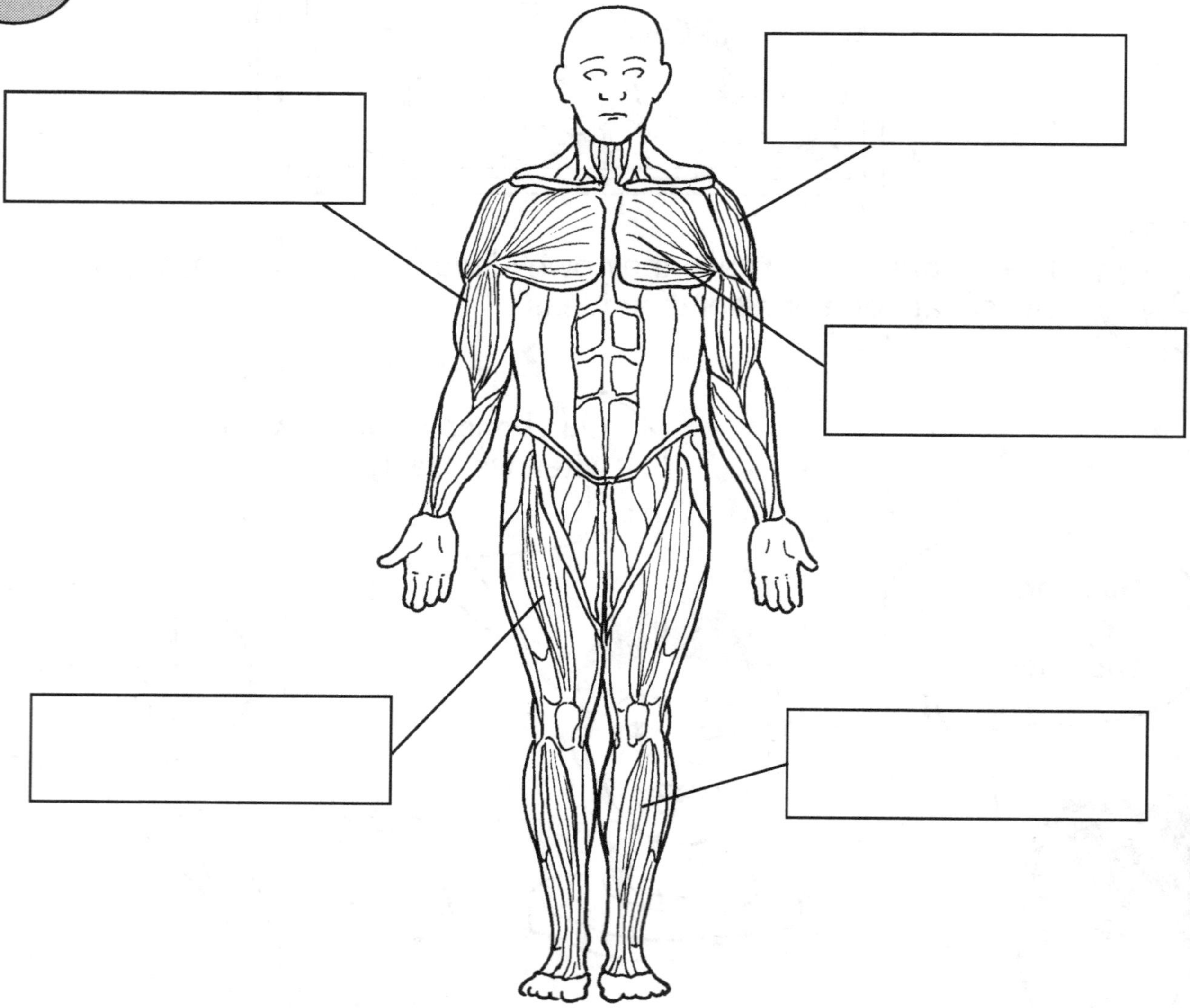

Useful words

Biceps   Pectoral   Deltoid
Tibialis   Quadriceps

Muscles and movement

Humans and Other Animals – Sheet 19

 Muscles usually work in pairs. When one contracts (**tightens**) the other relaxes (**loosens**). This means the movement is normally smooth.

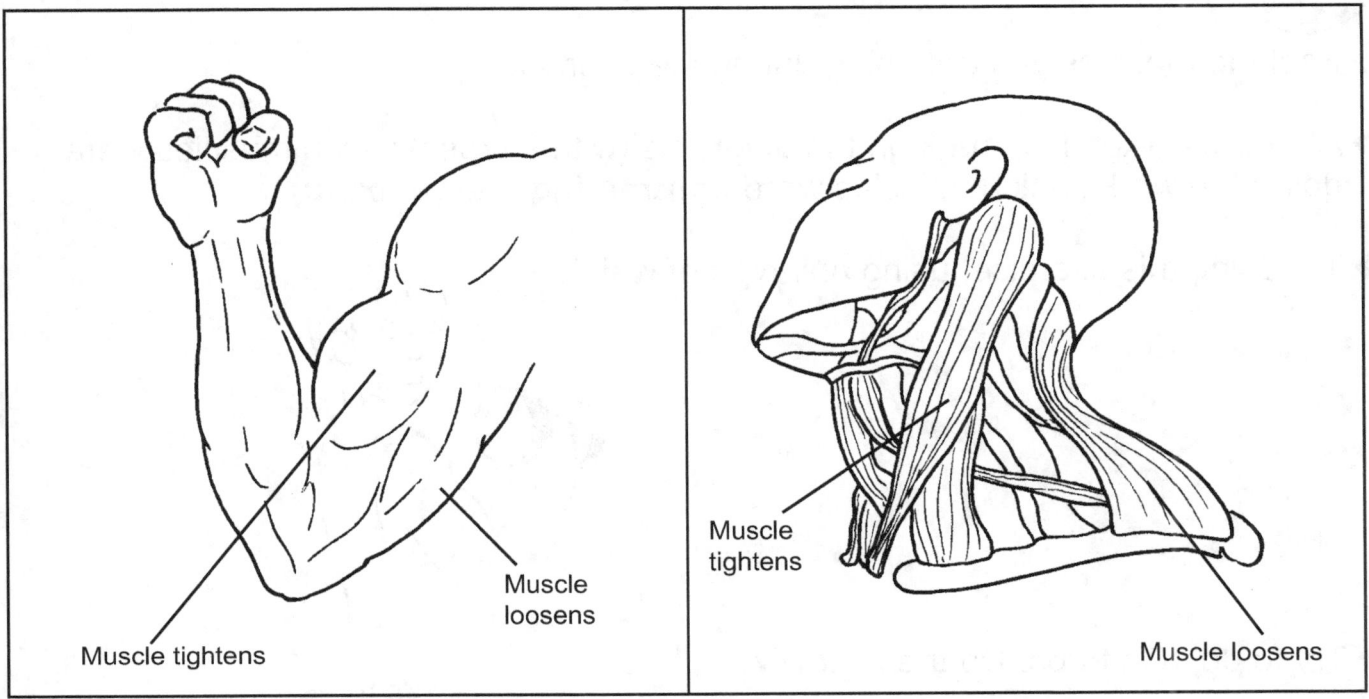

Muscle tightens

Muscle loosens

Muscle tightens

Muscle loosens

 Place your hand on the side of your neck and move your head slowly from side to side, then backwards and forwards.

 Describe what you feel when you move your muscles.

Working muscles

Humans and Other Animals
Sheet 20

You need: a weight.

 Try the following exercises which involve using different muscles in your body. Take care not to strain yourself!

◆ Feel the muscles you are using during the exercise.

◆ After each exercise, mark on the diagrams (with a cross) which muscles were doing the work (which muscles were **contracting** and **relaxing**).

◆ Try doing this exercise, using any type of weight.

◆ Try to do 10 sit-ups. Do them slowly.

◆ Sitting on your seat, put your feet flat on the floor. Slowly lift one heel (not your whole foot) up and down 10 times.

Life Processes and Living Things

© M Abraitis, A Deighan, B Gallaghar, B Smith & M Toner
This page may be photocopied for use by the purchasing institution only.

Green Plants

Background information

Growth and nutrition

Light is not needed for seeds to germinate but, once the shoot begins to grow leaves, light becomes essential for plant growth and survival through **photosynthesis** – the process by which a plant changes light energy into chemical energy (food). Living organisms are very sensitive to temperature changes. If the temperature of a plant's surroundings decreases, then biochemical reactions needed for life to exist begin to slow down. During the cold winter the growth of a tree (caused by cell division and enlargement) can stop altogether but when the warmer temperatures of spring arrive, the growing restarts. This starting and stopping of the tree's growth results in the formation of growth rings. The width of growth rings on a tree trunk gives a record of weather conditions when the tree was growing.

Water has many functions in the life of a plant. It is a necessary component of photosynthesis and is used for the transport of dissolved nutrients from the soil to the plant. A plant can increase in volume (grow) without the need to produce more cells. This occurs because water is absorbed by the plant's cells and makes them 'swell', thus increasing the overall size of the plant without a great increase in its dry weight.

The green leaves of a plant are essential for photosynthesis to occur. Carbon dioxide is absorbed through tiny holes in the leaf (called stomata) and water is taken up through the roots. The cells contain a chemical called chlorophyll in their chloroplasts which, with the help of sunlight, combines the water and carbon dioxide to form glucose and oxygen. The oxygen passes through the stomata in the leaves into the air and the glucose is used to provide the plant with energy for growth and reproduction. Some of the glucose joins together to form starch, which is stored in the cell's cytoplasm to form a food store. Other glucose molecules join into long chains called cellulose, which is used to construct the cell wall.

Helpful words
- Photosynthesis
- Nutrient
- Annular ring
- Cell
- Chloroplast
- Vacuole
- Nucleus
- Cytoplasm
- Stamen
- Carpel
- Ovary
- Pollen
- Stigma
- Fertilization
- Germinate

A simplified plant cell

Nucleus: contains the cell's DNA.
Chloroplast: is where photosynthesis occurs.
Vacuole: holds water and dissolved substances.
Cell wall: holds the cell together and keeps its shape.
Cytoplasm: many chemical reactions happen here.

Life Processes and Living Things

Green Plants

Photosynthesis, which occurs in the leaves of green plants, is the process by which both oxygen and food are produced to sustain life on Earth.

Chemical formula for photosynthesis

$$CO_2 + H_2O \longrightarrow C_6H_{12}O_6 + O_2 + \text{energy}$$

Carbon dioxide   Water   Light   Glucose   Oxygen

The roots of the plant, which are in the ground, are vital for absorbing water and nutrients, which are essential for the living cells of a plant to stay alive. The water passes up through the stem of the plant where it is transported to other parts of the plant. The water carries dissolved nutrients with it. The roots also help hold the plant in the ground, preventing it from being blown down easily.

Reproduction

Cross-pollination occurs when **pollen** from the top of one plant's **stamen** reaches the top, or **stigma**, of another plant's **carpel**.

The pollen contains the male sex cells and the **ovules** contain the female sex cells.

When the pollen lands on the stigma, it uses a sticky sugar substance present there to grow a pollen tube down into the plant's ovary where it meets an ovule. After **fertilization** (the combining of the male and female sex cells), a seed is produced which contains a young plant and a food store. The ovary can then develop into a fruit. The wall of the ovary can become soft and juicy, as in a cherry, or hard, as in a nut.

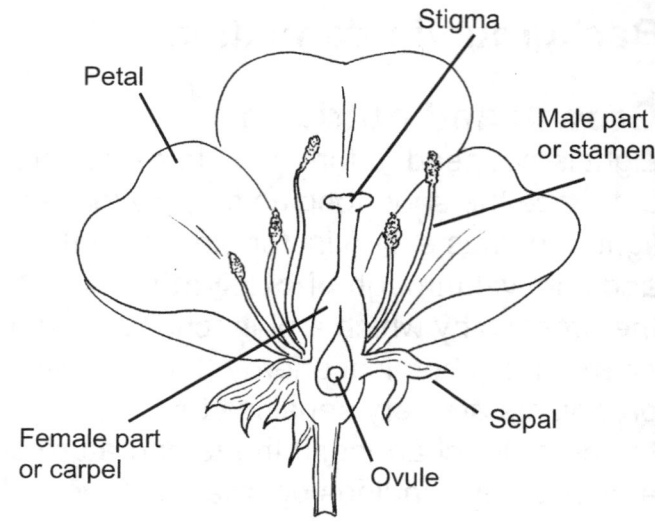

For a species of plant to survive, its seeds must be well dispersed to reduce competition between plants and to spread to other habitats. Some methods of dispersal are shown below:

Sycamore: Wind

Apple: Animal

Coconut: Water

Pea: Bursting

An edible fruit encourages animals to eat it. The animal will either spit out the seed or swallow it: the seed's tough coating prevents it from being digested and allows it to pass through the animal, to be dropped far away from the parent plant unharmed.

Green Plants

When a seed is scattered, it has little water in it. For the seed to **germinate**, or grow into a new plant, it must have water, oxygen and a suitable temperature. If the conditions are right, the seed uses its food store to grow a new root to take in water, and then a new shoot will begin to grow.

Some ideas for further work
- Find out if there is a link between hay fever and the amount of pollen in the air.

Life Processes and Living Things

Green Plants

Activity	Answers	Resources needed	Teaching/safety notes
Sheet 1 (page 48) Plants need light	The plant in the closed box will wilt, and possibly die.	3 similar plants 2 cardboard boxes	Photosynthesis cannot occur without light. Insufficient light will cause retardation of growth and possible death of the plants.
Sheet 2 (page 49) Plants need water	Plants will not thrive with too much or too little water.	4 cress seedlings Ruler Measuring cylinder	All plants require water because it contains dissolved nutrients and because it is needed for reactions such as photosynthesis. Water is normally taken up through the roots. These roots also require some air. As water soaks through the soil it draws in air. If the soil becomes water-logged, then no air is drawn in and this will affect the plant's growth.
Sheet 3 (page 50) Plants need warmth	The plants in the warm soil should grow better.	2 trays of cress seedlings Thermometer String Thick piece of transparent plastic	
Sheet 4 (page 51) Plants and leaves	The plant with more leaves should show the greater increase in growth over the four weeks.	Ruler 2 geranium plants	The process of photosynthesis occurs in the leaves of plants.
Sheet 5 (page 52) Roots		Jar Paper towel Broad bean, soaked overnight	Seeds require water, oxygen (in the air) and a suitable temperature for germination. As the young plant grows, it digests the food stored in the seed for energy and growth. Once the leaves grow, the plant can make its own food by photosynthesis. The root will grow downwards and the shoot will grow upwards. It is thought that this growth is responding to the Earth's gravity.

Life Processes and Living Things

Green Plants

Activity	Answers	Resources needed	Teaching/ safety notes
Sheet 6 (page 53) Stems	The dye should be present up through the length of the stem. This shows that water travels up through the stem of a plant to all its other parts.	Stick of celery Jar Red food colouring Hand lens	This activity could also be carried out with carnations and with different colours of food dye.
Sheet 7 (page 54) Flowering plants and seeds	1 Cherry tree, cherry, seed 2 Peach tree, peach, seed 3 Apple tree, apple, seed		
Sheet 8 (page 55) Dispersal of seeds	Peas – Bursting Apples – Animal Coconuts – Water Sycamore – Wind Cherries – Animal Dandelion – Wind	Reference materials	Some seeds have juicy fruits around them to encourage animals to eat them. The seeds could be spat out by the animal or swallowed. The hard coatings prevent the seeds from being digested and they pass out of the animal. Both situations will result in the plant's seeds being dispersed over a larger area.
Sheet 9 (page 56) Pollination	Under low magnification pollen looks like small round particles of dust.	Daffodil Low-power microscope	
Sheet 10 (pages 57 and 58) Parts of the flower		Coloured pencils Scissors Glue Card	
Sheet 11 (page 59) Light and germination	Light is not required for germination (seeds are normally planted underground).	Cress seeds Paper towels 2 margarine tubs	The young plants which were germinated in the dark will be tall, thin and pale as they will not be able to photosynthesize. Once the food store is used up, they will die.

Life Processes and Living Things

Green Plants Sheet 1

Plants need light

You need: 3 similar plants and 2 cardboard boxes.

 **Read**

Green plants need light, water and a suitable temperature to grow well. Gardeners sometimes use greenhouses to keep their plants warm and to protect them from frost.

 **Activity**

◆ Investigate the effect of light on the healthy growth of plants by putting three plants on the same window ledge.

◆ Leave one plant on its own, put one in an open cardboard box and another in a closed cardboard box.

 **Answer**

After a week, describe how each of the three plants look.

Life Processes and Living Things

Plants need water

Green Plants
Sheet 2

You need: 4 cress seedlings, a ruler and a measuring cylinder.

- Investigate whether too much or too little water affects the growth of cress seedlings.

- Use a measuring cylinder to water the cress seedlings by different amounts every two days.

- Use a ruler to measure the height of the seedlings before you water them.

Write in the height of the seedlings every 2 days	No water	5 cm³ of water each time	20 cm³ of water each time	50 cm³ of water each time
At the beginning of the investigation				
After 2 days				
After 4 days				
After 6 days				
After 8 days				
After 10 days				

Does too much or too little water affect the growth of the seedlings?

Green Plants Sheet 3

Plants need warmth

You need: 2 trays of seedlings, a thermometer, string and a thick piece of transparent plastic.

- Investigate the effect of warmth on the growth of cress seedlings.

- Take two trays of cress seedlings. Keep one of the trays of seedlings warm by folding a piece of thick transparent plastic over it.

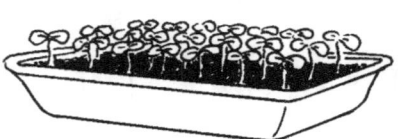 

◆ Put both trays in the same sunny place and measure the temperature of the soil every two days.

Write in the temperature of the soil every 2 days	At the start	After 2 days	After 4 days	After 6 days	After 8 days	After 10 days

Does covering up the plants to keep them warm help them grow better?

Life Processes and Living Things

Plants and leaves

Green Plants — Sheet 4

You need: a ruler and 2 geranium plants.

 Living plants have three main parts: roots, stem and leaves.

 Label the three main parts of this plant.

Now let's investigate whether leaves help a plant grow.

◆ Take two geraniums of the same size.

◆ Take half the leaves off one of the geraniums and then place them both in the same part of the room.

Plant 1

Plant 2

◆ Measure the height of each plant every week.

	Week 1	Week 2	Week 3	Week 4
Height of plant 1				
Height of plant 2				

Remember, water the plants equally and keep them in the same place.

© M Abraitis, A Deighan, B Gallagher, B Smith & M Toner
This page may be photocopied for use by the purchasing institution only.

Life Processes and Living Things

Green Plants
Sheet 5

Roots

You need: a jar, paper towel, some water and a broad bean which has been soaked overnight.

 **Read**

Plant roots have two jobs. One is to fix the plant in the ground so that it does not fall over easily. The other job is to soak up water from the ground into the stem of the plant.

The roots of a tree and of a seedling have the same purpose. They fix the plants to the ground and they soak up water.

Activity Use the experiment below to investigate how the roots of a broad bean begin to grow.

Roll a paper towel into a cylinder.

Put the towel into a jar and add some water.

Push a broad bean seed between the towel and the jar.

Sketch the bean every two days.

At the start	Day 2	Day 4	Day 6	Day 8	Day 10

Life Processes and Living Things

Stems

Green Plants
Sheet 6

> You need: a stick of celery, a jar, red food colouring and a hand lens.

   Investigate the stem of a plant by doing the following activity.

◆ Add some red food dye to a beaker or jar of water.

◆ Place a stick of celery into the coloured water and leave it for three or four days.

◆ Ask your teacher to slice the celery stick in two, lengthwise, and then examine it with a hand lens.

   ◆ Draw what you see and explain what has happened.

Flowering plants and seeds

Green Plants — Sheet 7

Read

'From tiny acorns mighty oak trees grow.'

Many plants produce seeds which, when dropped into the soil in the right conditions, grow into new plants.

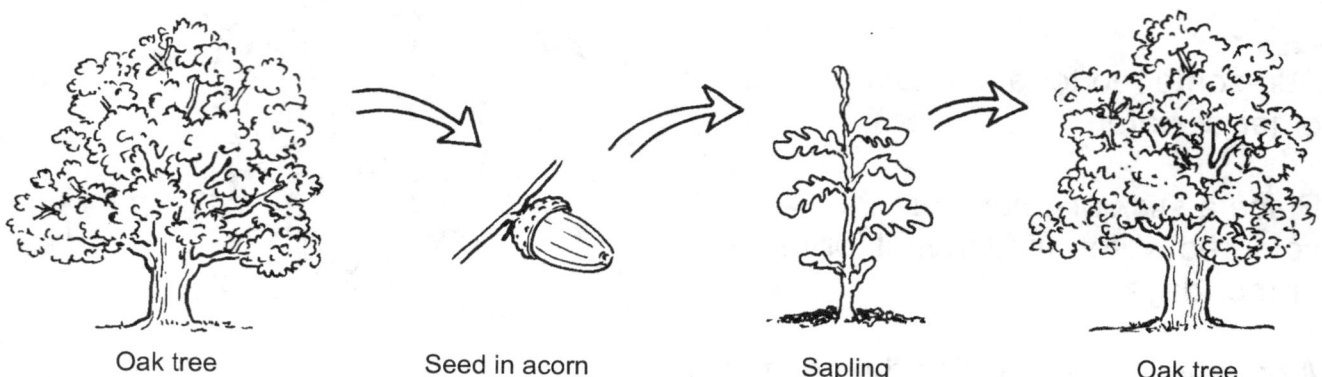

Oak tree · Seed in acorn · Sapling · Oak tree

Activity

Use the words to name the plants and their fruits below. The pictures show the plants, their flower, the fruits and the seeds inside the fruits.

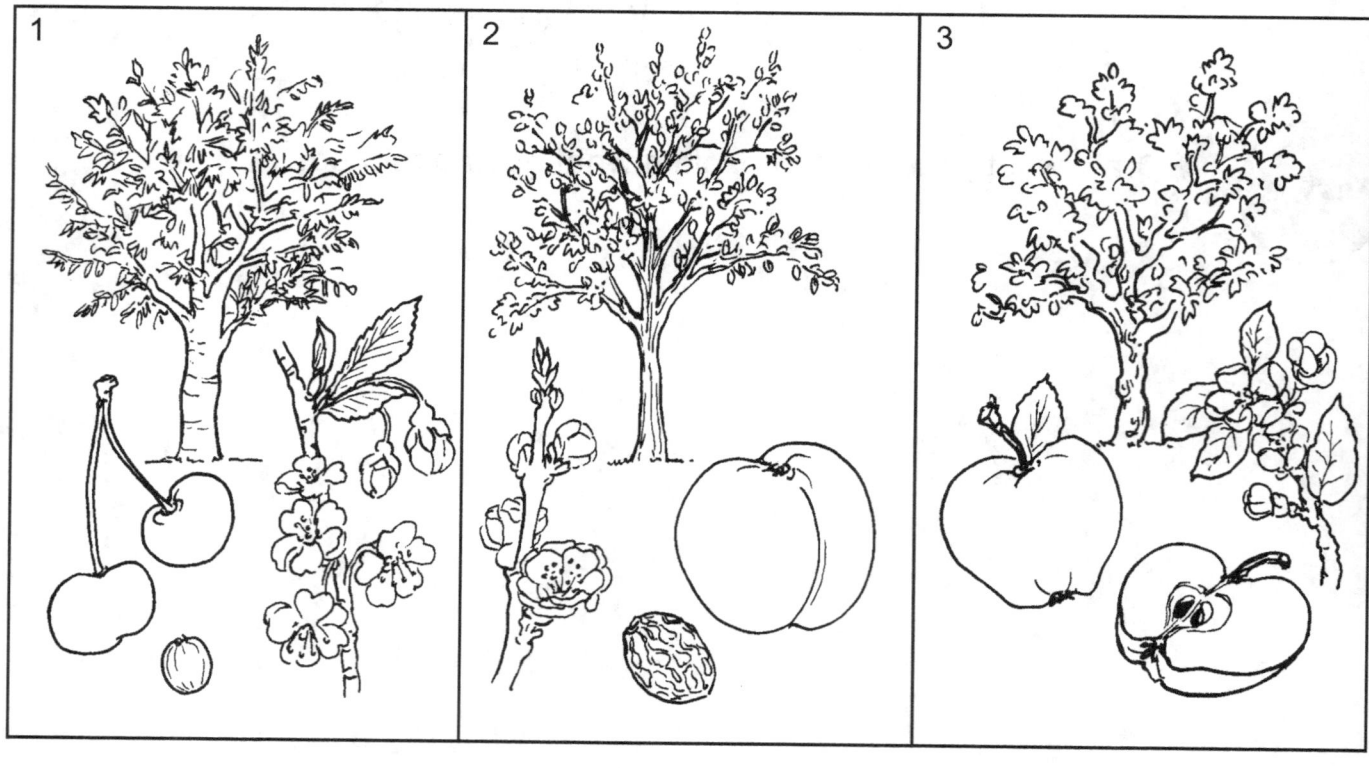

Apple tree   Peach tree   Cherry tree   Apple

Peach   Cherry   Seed

Life Processes and Living Things

Dispersal of seeds

Green Plants
Sheet 8

 Write the correct method of dispersal beside each seed.

Peas Apples Coconuts

Methods of dispersal
Animal   Water   Wind   Bursting

 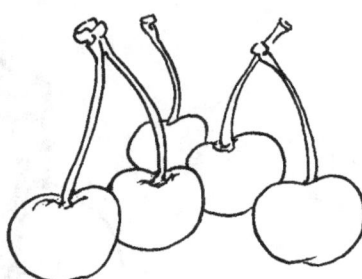 

Sycamore Cherries Dandelion

 Why do seeds which rely on animals to scatter them have juicy fruits around them?

© M Abraitis, A Deighan, B Gallagher, B Smith & M Toner
This page may be photocopied for use by the purchasing institution only.

Life Processes and Living Things

Pollination

Green Plants
Sheet 9

You need: a daffodil and low-power microscope.

 **Read** Before a plant can produce fruits and seeds it must be pollinated by pollen from another plant. The pollen cannot move on its own but has to be carried by insects or the wind.

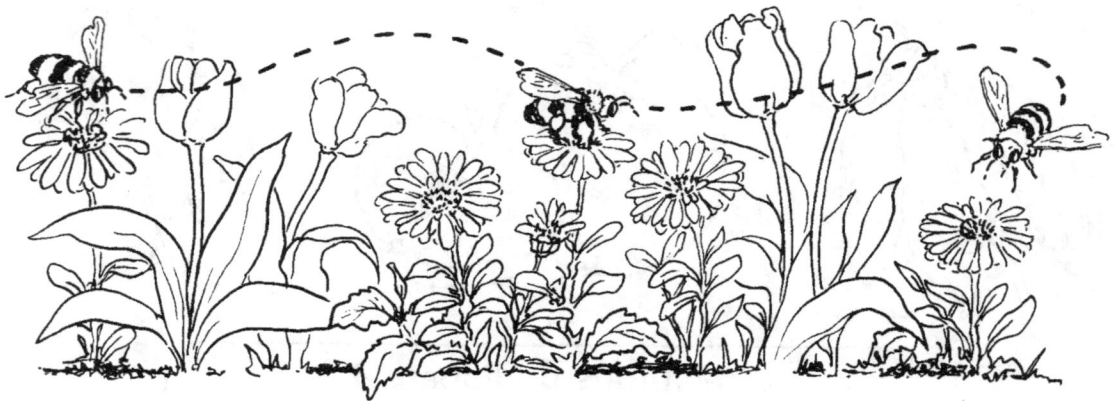

 **Activity** You can study pollen by sprinkling some from a daffodil onto a microscope slide and then looking at it under a microscope.

 **Answer** Draw what the pollen looks like under the microscope.

Parts of the flower

Green Plants
Sheet 10

You need: coloured pencils, scissors, glue and card.

 **Read** To reproduce themselves, plants need both a male and a female part. Both the male and female parts of a plant are contained in a **flower**.

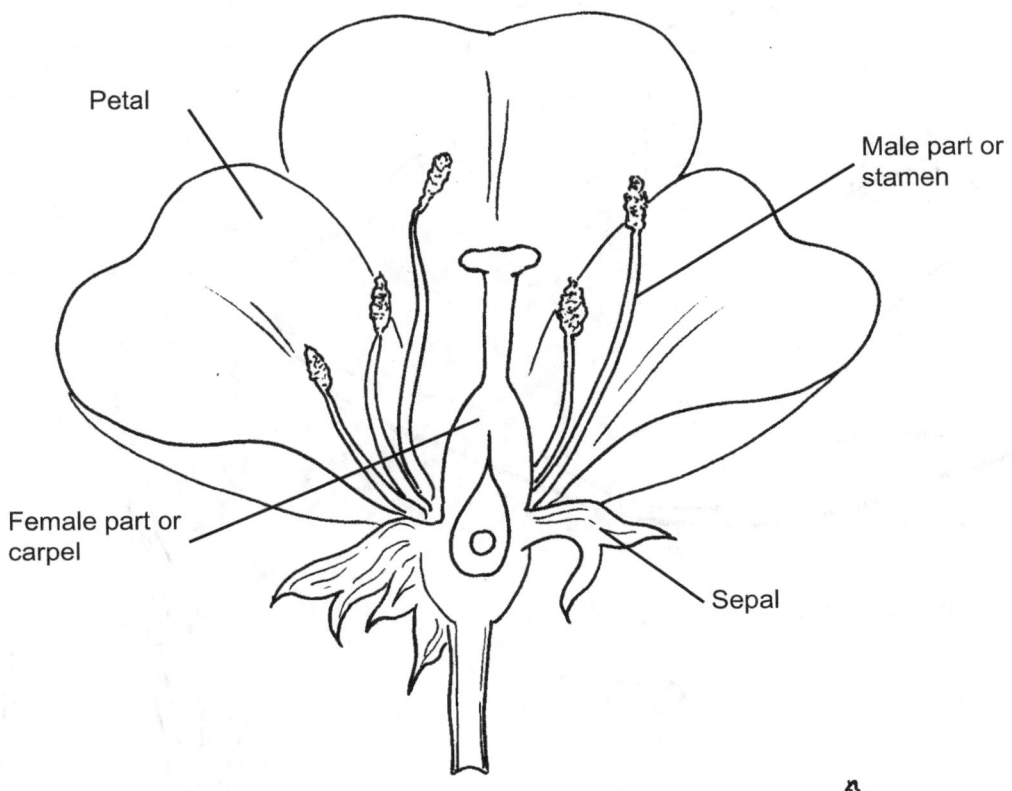

The male cells, which are called **pollen**, are made in the **stamen**.

The female cells, which are called **ovules** (or eggs), are made in the **carpel**.

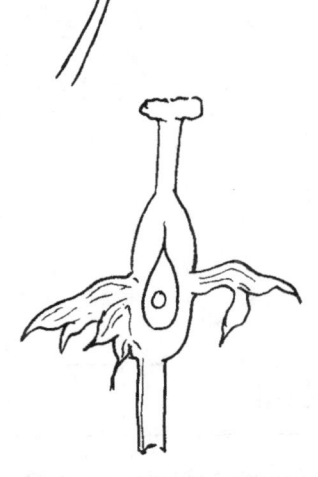

Parts of the flower

Green Plants
Sheet 10 – Continued

- Colour the parts of the flower below.
- Glue them to a piece of card, then cut them out to make a flower.
- Use the names below to label your flower.

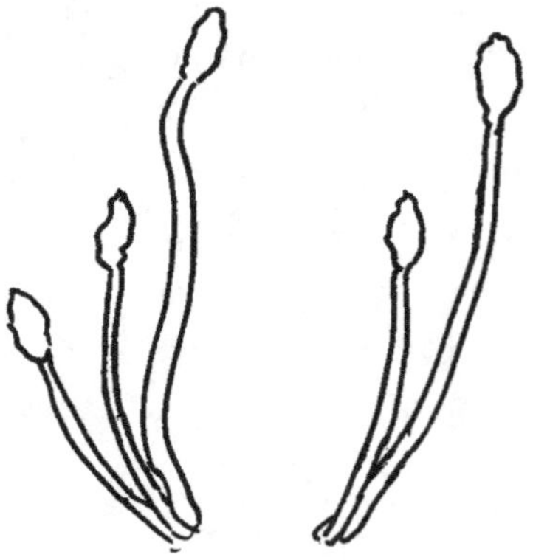

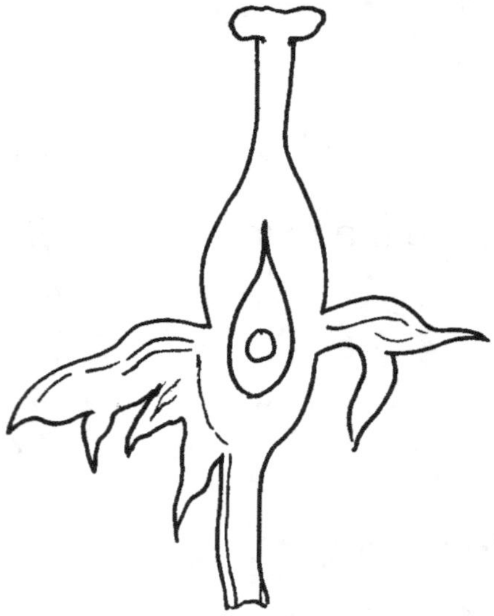

| Petal | Carpel | Sepal | Stamen |

Life Processes and Living Things

Light and germination

Green Plants
Sheet 11

| You need: cress seeds, paper towels and 2 margarine tubs. |

 **Read**

The right conditions are necessary for a seed to **germinate**, or grow into a new plant.

 **Activity**

Investigate whether light is needed for a seed to germinate.

♦ Put some paper towels on the bottom of two margarine tubs and water them equally.

♦ Sprinkle the same amount of cress seeds (10 to 20 seeds) onto the paper towels.

♦ Punch some holes in one of the lids (to let air in) and put it on one of the margarine tubs so that the seeds are in the dark. Leave the other tub uncovered.

♦ Leave the two tubs in a warm, sunny place for a week or two.

 **Answer**

Is light needed for seeds to germinate?

 **Look further**

Devise your own investigations to see whether warmth and water are needed for seeds to germinate.

© M Abraitis, A Deighan, B Gallagher, B Smith & M Toner
This page may be photocopied for use by the purchasing institution only.

Life Processes and Living Things

Variation and Classification

Living things are **classified** by groups according to their similarities. For example, we can classify birds together because of their feathers, or fish because of their wet scales and fins.

Helpful words
- Classify
- Key
- Vertebrate
- Invertebrate
- Mammal

A simple **key** for assigning animals to groups is shown below. For animals, the first distinction made is between those animals which have a backbone (**vertebrates**) and those which do not (**invertebrates**). The vertebrates are separated into five other groups depending on their features. **Mammals** have hair and supply milk to their young, **birds** have feathers and lay hard eggs, **fish** have wet scales, gills and fins, **reptiles** have dry scales and lay soft eggs, and **amphibians** have moist skin and live in a land and water habitat. Invertebrates can also be separated into further groups such as **insects**, **arachnids**, **crustaceans**, **myriapods** and others.

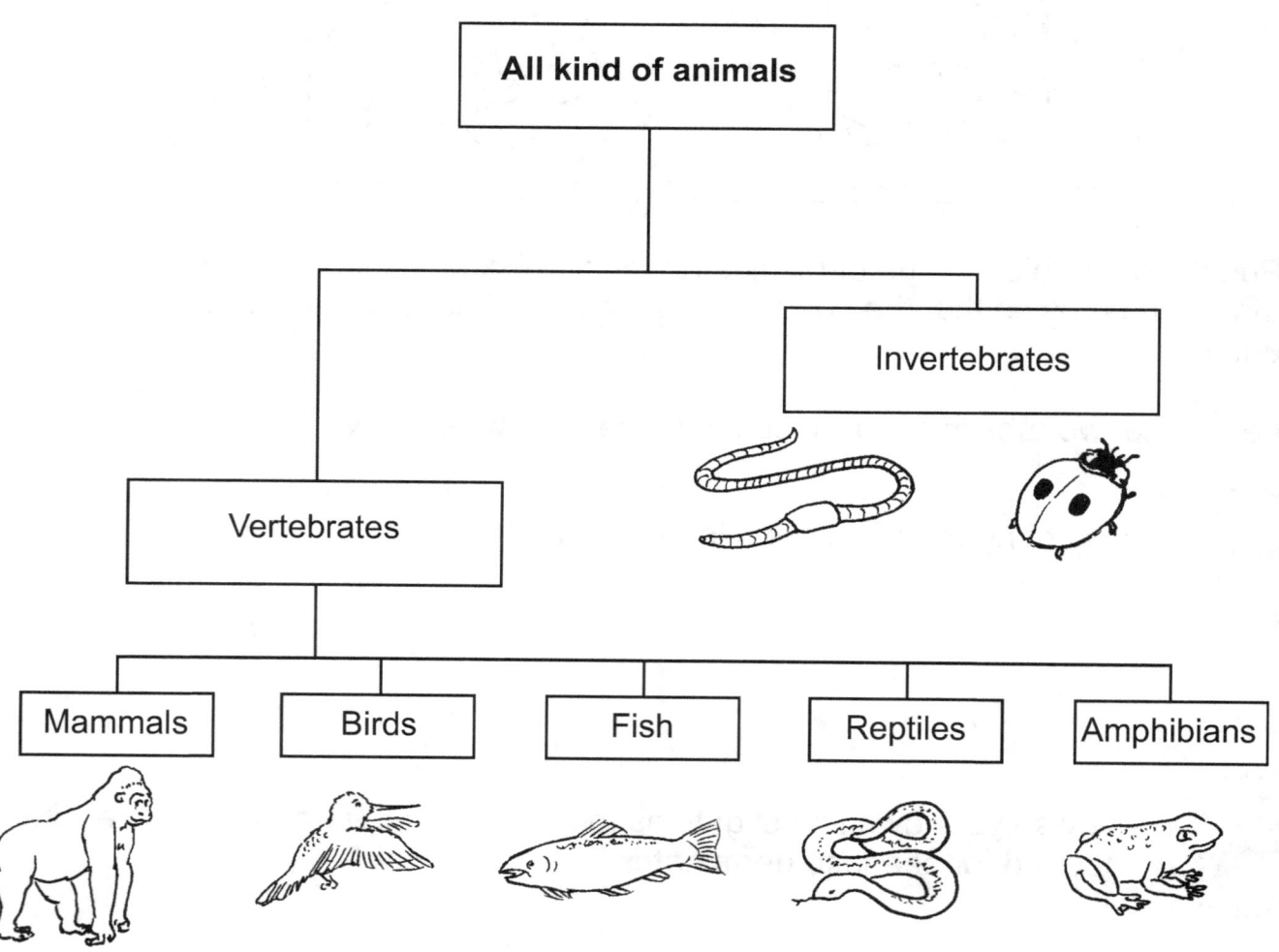

Life Processes and Living Things

Variation and Classification

Plants are also classified into groups according to their characteristics. Examples include **algae**, **fungi**, **mosses**, **ferns** and **seed plants**. Plants that produce seeds are classified as **flowering** and **non-flowering** plants (cone plants).

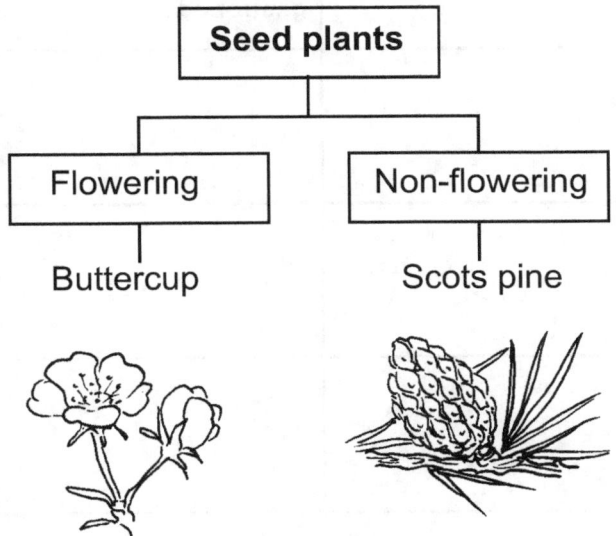

Some ideas for further work

◆ Make up a key which can be used to identify the other children in your group or class.

◆ Draw your own imaginary creatures (monsters or space creatures), give them names and then make up a key so that they can be identified.

Life Processes and Living Things

Variation and Classification

Activity	Answers	Resources needed	Teaching/ safety notes
Sheet 1 (page 63) Groups of living things	The children could sort by: • wings and legs (1, 5, 7, 13) • 8 legs (4, 9, 15) • shell (3, 6, 16) • no legs (2, 8, 11) • 6 legs (10, 12, 14).	Scissors Glue Reference materials	Spiders have 8 legs and are called **arachnids**. **Insects** such as ants, termites and cockroaches have 6 legs. Segmented worms are **annelids** and shelled snails are **gastropods**.
Sheet 2 (page 64) Mini-beast key	A = Worm B = Ant C = Bee D = Snail		
Sheet 3 (page 65) Naming mini-beasts	A = Spider B = Wasp C = Honey bee D = Ant		
Sheet 4 (page 66) Dog key	A = Benjy B = Droopy C = Cassie D = Rover		
Sheet 5 (page 67) Big cat key	A = Panther B = Lion C = Leopard D = Tiger		
Sheet 6 (page 68) Prehistoric animal key	Top left = Wings Top right = No wings Middle, from left to right: Tail, No tail, Horns, No horns Rhamphorhynchus = A Pterodactyl = D Triceratops = C Tyrannosaurus rex = B	Reference materials	

Life Processes and Living Things

Living things

Variation and Classification — Sheet 1

You need: scissors and glue.

 Living things can be grouped together by their similarities. For example, birds can be grouped together because they all have feathers and lay hard eggs.

- Use words like **wings and legs**, **shell**, **no legs**, **six legs** and **eight legs** to group the animals below into similar groups.

- Cut out the animals and sort them into similar groups.

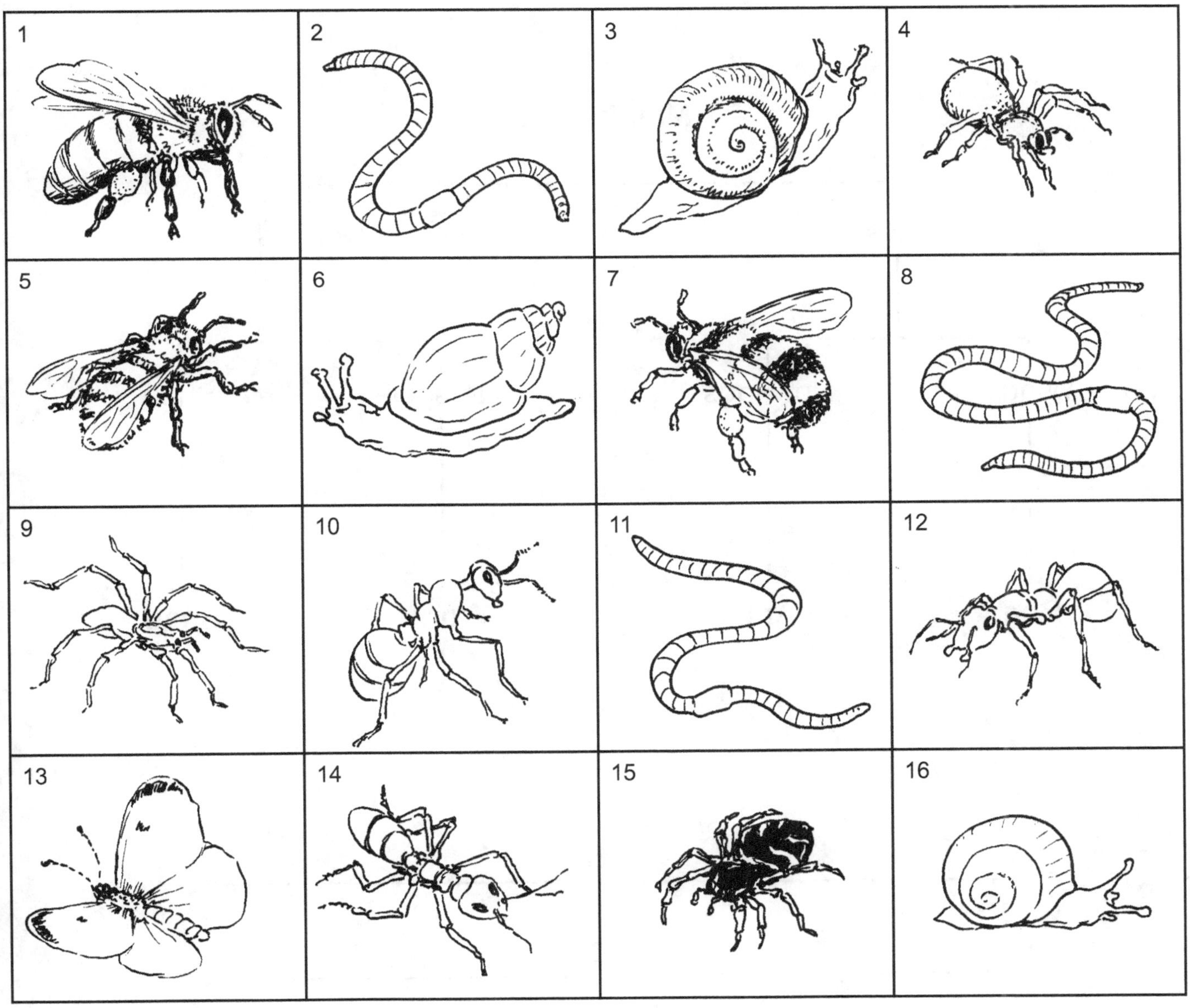

1 bee	2 worm	3 snail	4 spider
5 bee	6 snail	7 bee	8 worm
9 spider	10 ant	11 worm	12 ant
13 butterfly	14 ant	15 spider	16 snail

© M Abraitis, A Deighan, B Gallagher, B Smith & M Toner
This page may be photocopied for use by the purchasing institution only.

Variation and Classification
Sheet 2

Mini-beast key

 **Activity** Use the key below to find the names of the mini-beasts labelled A, B, C and D.

A   B   C   D

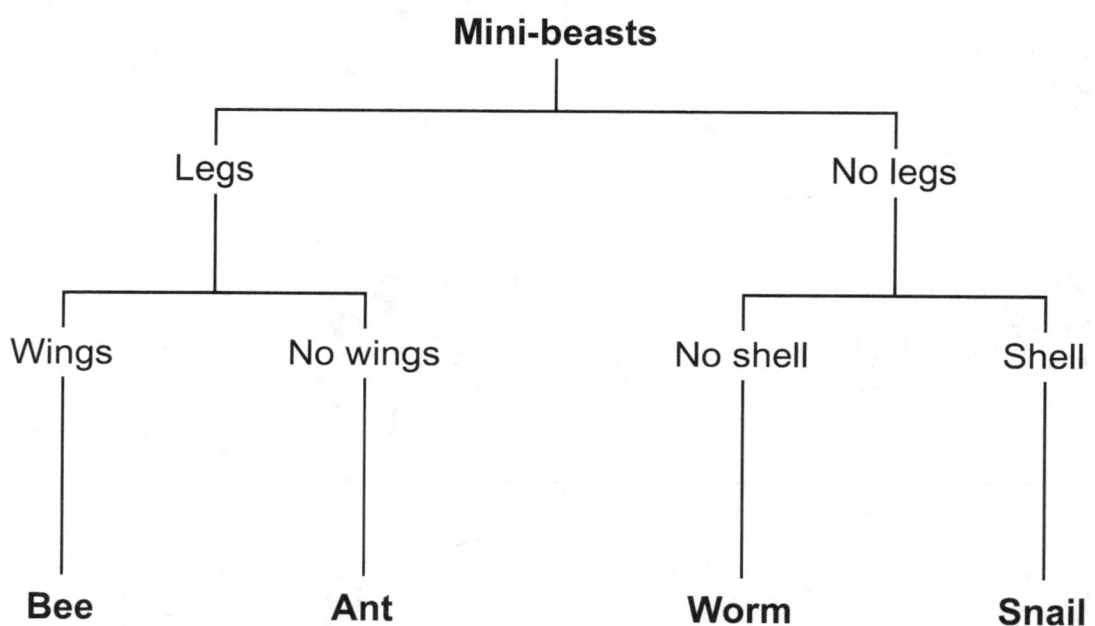

Answer A _____   B _____

C _____   D _____

Life Processes and Living Things

Variation and Classification
Sheet 3

Naming mini-beasts

 **Activity** Use the key below to find the names of the mini-beasts labelled A, B, C and D.

A B C D

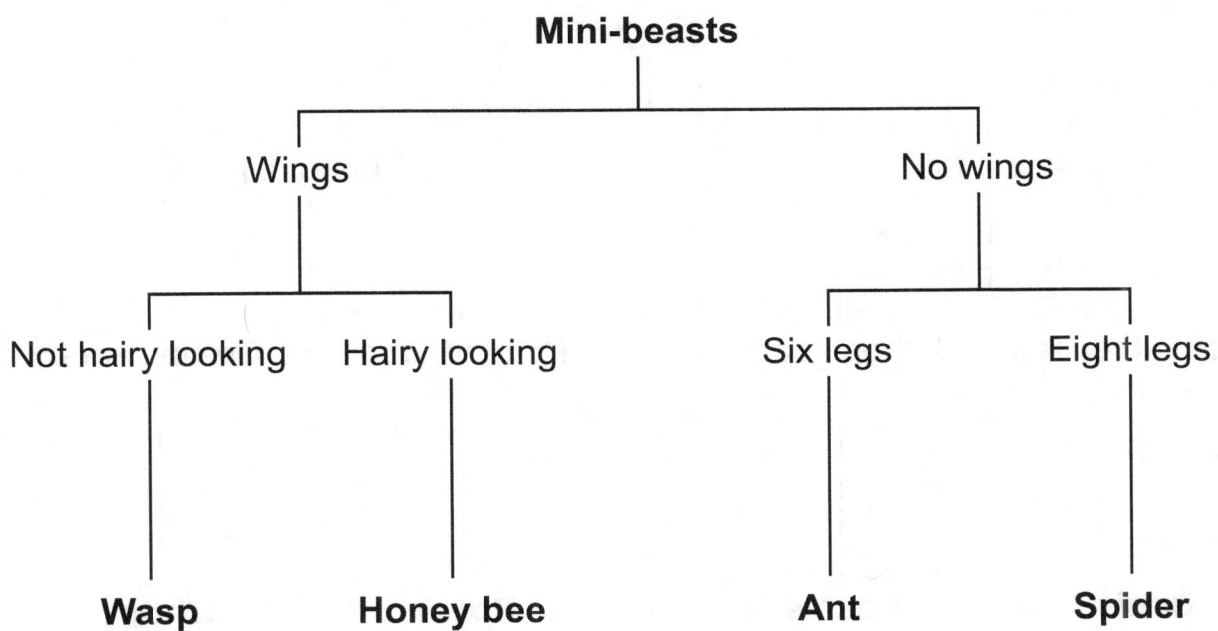

Answer

A _____ B _____

C _____ D _____

Variation and Classification
Sheet 4

Dog key

 **Activity** Use the key below to find the names of the dogs labelled A, B, C and D.

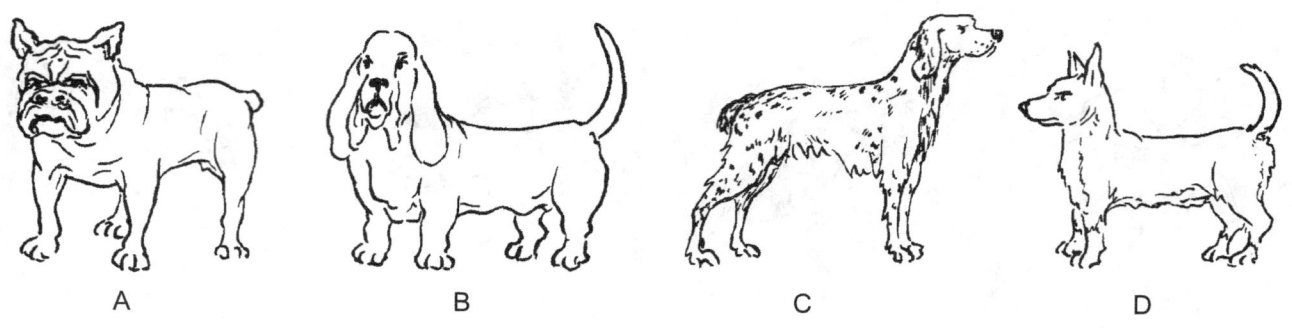

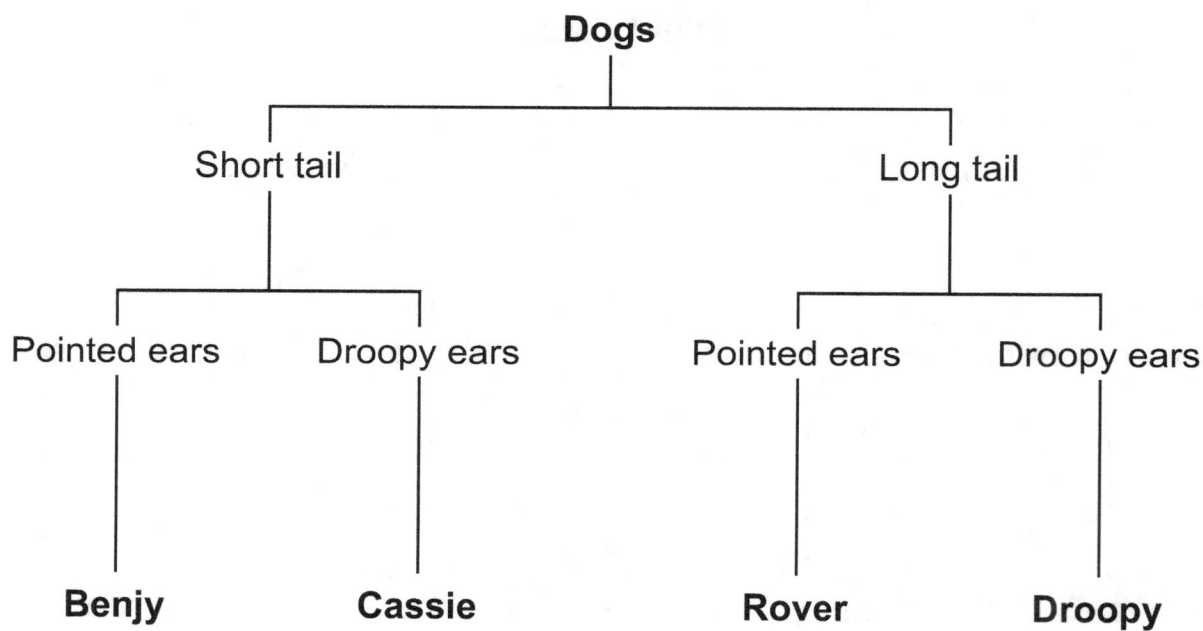

Answer   A _____   B _____

C _____   D _____

Life Processes and Living Things

Big cat key

Variation and Classification
Sheet 5

Activity Use the key below to find the names of the cats labelled A, B, C and D.

A B C D

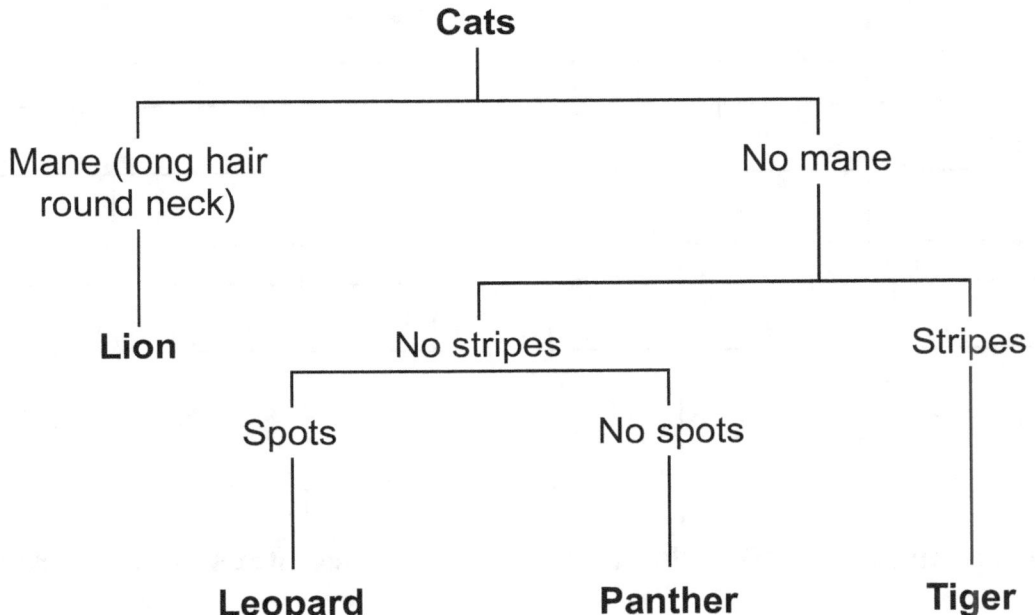

Answer

A _____ B _____

C _____ D _____

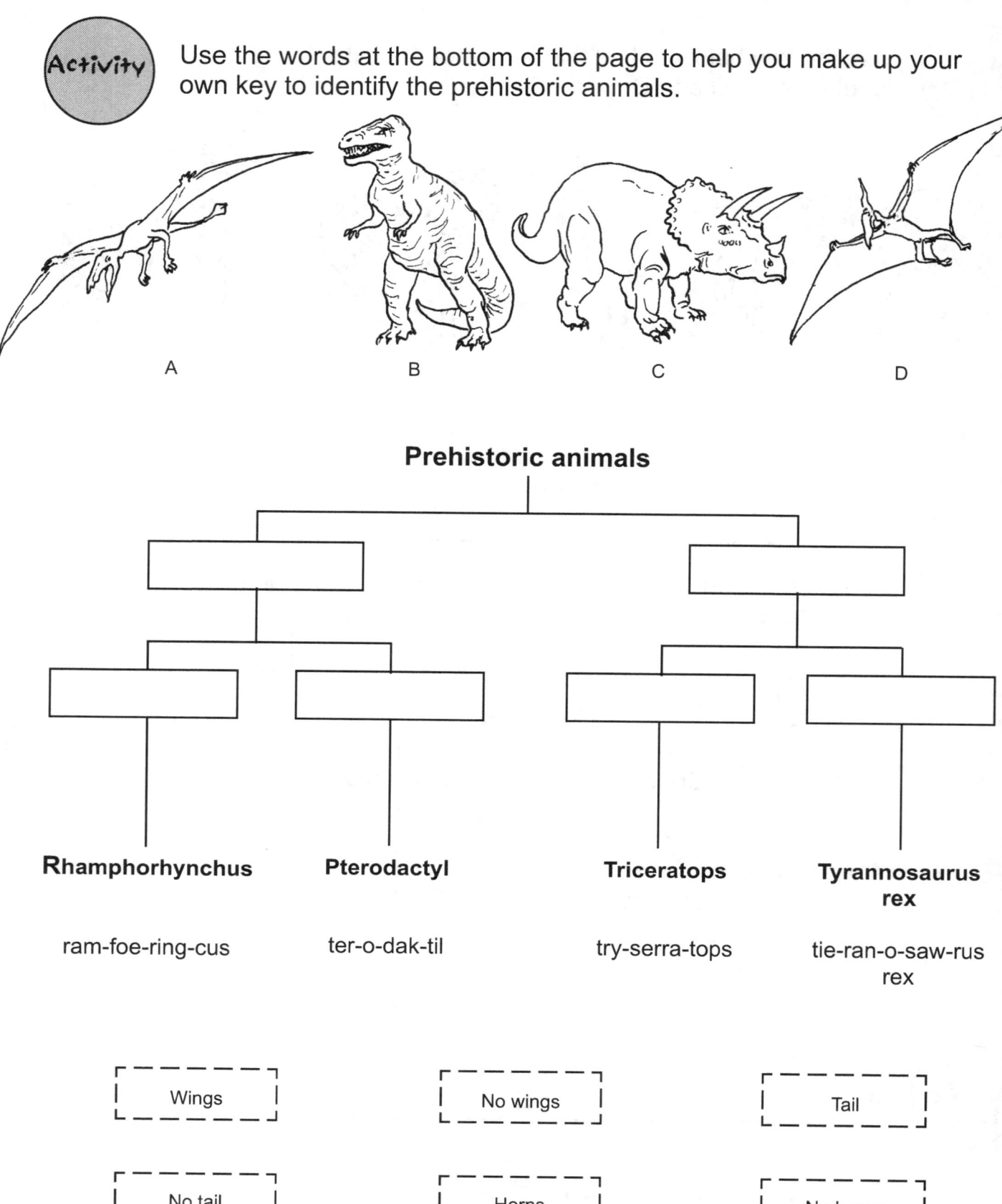

Living Things in their Environment

Background information

Adaptation

Animals and plants live in an **environment** which supplies all their needs, such as food, water and shelter, and to which they are best adapted to live. Ferns are plants which prefer damp, shady conditions in which to live. Ferns are not found everywhere because not every environment is damp and shady. Ferns are often found living in woodlands. Squirrels are animals which also prefer a woodland environment as it supplies them with all their needs, such as trees to live in and acorns on which to feed. Squirrels are suited to this environment as they have sharp claws for climbing and tails for balancing.

A pond or river will have plants and animals in or around it which are best suited to that environment. Otters live on river banks, where they can burrow for shelter and feed on fish. An otter is suited to this environment as it has webbed feet and a waterproof coat. Water lilies are suited to their environment as they can receive more sunlight by floating on the surface of the water.

Where a group of living organisms lives is called their **habitat**. The habitat provides the organisms (collectively known as a **community**) with food, water and shelter.

Helpful words	
Environment	Food web
Habitat	Predator
Microbe	Prey
Ecosystem	Virus
Food chain	Bacteri
Producer	Fungus
Consumer	Protozoa

The community of organisms, the habitat and the environmental factors which affect them, such as the weather, are collectively called an **ecosystem**.

Feeding relationships

Animals and plants in an ecosystem are dependent on each other. For example, a squirrel relies on trees for shelter and a daffodil relies on insects for pollination. Living organisms in an ecosystem are also dependent on each other for food. This dependency can be described as a **food chain**. A food chain normally begins with a green plant. The green plant makes food from the Sun and is called a **producer**. The animals which eat the plant are called **primary consumers** and the animals which eat them are called **secondary consumers**. Recently, however, organisms have been discovered in the deep ocean which feed on sulphur found near hot volcanic outcrops. They are dependent on neither sunlight nor green plants as their source of food.

A food chain

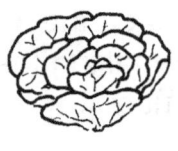 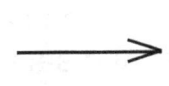  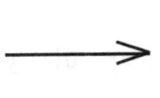 

Producer → Eaten by → Primary Consumer → Eaten by → Secondary Consumer

Life Processes and Living Things

Living Things in their Environment

A simplified food web

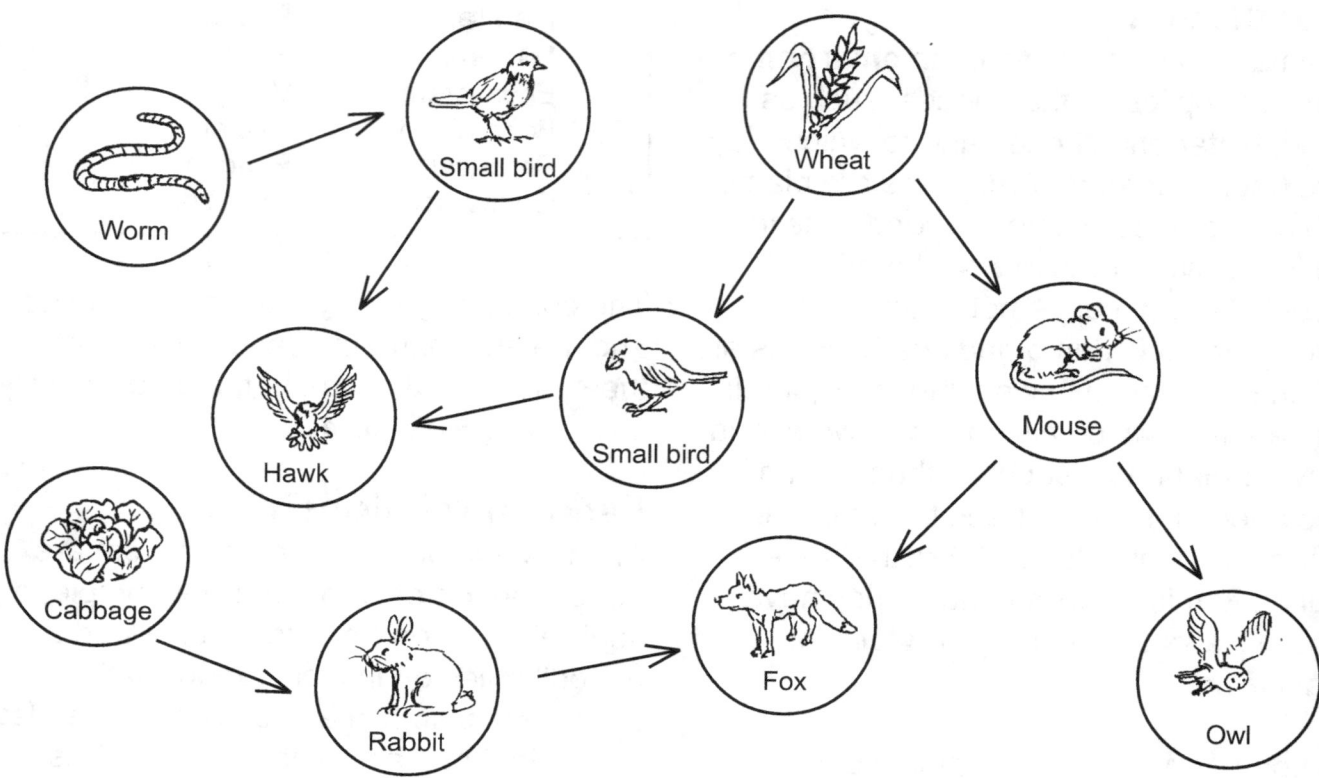

Food webs show the feeding relationship between many organisms as a collection of food chains. A food web is also a good way of illustrating how dependent living things are on each other for food in an ecosystem.

Micro-organisms

Microbes are very small living organisms and cannot be seen by the naked eye. Due to the differences between microbes and other living things, they are classified in a group of their own. There are four main types of microbe: **viruses**, **bacteria**, **fungi** and **protozoa**.

The smallest microbes are viruses and they cannot reproduce by themselves. They are parasites and rely on invading animals and plant cells, taking control of them to make them produce more viruses. Viruses cause many diseases, such as flu, chicken pox, cold sores and HIV.

Bacteria look like living cells but they do not have a nucleus. Certain bacteria are responsible for making foods rot and they are also the cause of some diseases, such as cholera, diarrhoea, leprosy and diptheria. Some bacteria live in the human intestines (such as E coli) and are also present in our mouths. Bacteria are used in sewage works. Bacteria feed on the organic matter in the sewage, making the water cleaner and safer. One useful

Living Things in their Environment

by-product of this is methane gas, which can be used as fuel.

Fungi are made from long thread-like structures called hypha. A group of hyphae is called a mycellium. Fungi can cause diseases such as athlete's foot (dermatophytosis interdigitale), but they have useful functions as well. Yeast is a fungus and is used in making bread. Fungi also help decompose the flesh of dead organisms and release the nutrients back into the soil. Plants reuse these nutrients and the cycle begins again.

Protozoa look like individual cells and can also cause disease. Malaria is a disease which is caused by protozoa living in a human's red blood cells. It is transmitted by the female mosquito.

Some ideas for further work

- Devise your own ecosystem. Remember that most organisms (living things) need water and air to survive.

In the school grounds create a small shallow pond. Surround it with vegetation, both living and dead (plants, grass, tree bark and plant cuttings, etc) and small stones. Leave it undisturbed for a month and from then on keep regular notes on any changes that you observe. These changes should include animal or vegetation changes.

- Find out who the scientist Charles Darwin was and something about what he studied.

- Find out about Louis Pasteur and what he found out about microbes.

- About one quarter of all the deaths in the world are due to diseases caused by microbes. Find out ways in which microbes can spread from person to person and ways in which this spread can be prevented.

Living Things in their Environment

Activity	Answers	Resources needed	Teaching/safety notes
Sheet 1 (page 76) Habitats	• Woodland: squirrel, fox, badger, owl. • Field: hare, mouse, horse, dairy cow. • Pond: goose, frog, tadpole, perch.	Scissors Glue Reference materials	
Sheet 2 (page 77) Which habitat?		Reference materials	The children could use other resources to research suitable habitat(s) for each of the animals.
Sheet 3 (page 78) Which habitat for woodlice?	The woodlice prefer damp and dark areas, so there should be more of them under the stones and wet bark.	Tray with sand Dry leaves Stones Wet bark Woodlice	Return the woodlice to where you found them when the investigation is finished.
Sheet 4 (page 79) The wormery	The worms burrow in the ground. They pull underground dead leaves from the top of the soil.	Wormery 6 worms Leaves Cardboard	When the worms burrow they break the soil into finer parts and pull leaves into the ground. They decay and enrich the soil. Their burrows help air and water to reach the roots of plants and, as they move through the soil, they bring new soil to the top. Return the worms to where you found them when the investigation is finished.
Sheet 5 (page 80) Animals in their habitat 1	1 The bird is feeding on a worm. 2 The frog is catching an insect. 3 The eagle has caught a fish. 4 The owl has caught a mouse. 5 The koala bear is feeding on eucalyptus leaves. 6 The cow is grazing on grass. 7 The mouse is eating grains of wheat. 8 The pike is chasing the perch (small fish). 9 The panda is eating bamboo.	Reference materials	

Living Things in their Environment

Activity	Answers	Resources needed	Teaching/ safety notes
Sheet 6 (page 81) Animals in their habitat 2	• The goose has: waterproof feathers, beak for gripping food, webbed feet for swimming. • The squirrel has: tail for balance, strong teeth for biting hard, sharp claws for gripping.	Reference materials	
Sheet 7 (page 82) Adaptation	• The polar bear's white fur will camouflage it against the snow. Its sharp claws and teeth will help it catch and kill its prey. Its good eyesight will help it see its prey. Its claws and fur on its soles will prevent it from slipping and its blubber will keep it warm when it swims between the ice floes. • The camel can survive the journeys between the water spots and its splayed feet will help it walk on the sand. Its hairy eyelashes and nostrils will prevent sand from being blown into them and its coat will help keep it warm at night.	Reference materials	There are many other examples of animals and plants which have adapted to live in extreme environments, such as penguins and cactii.
Sheet 8 (page 83) What is soil?	Any large stones should sink to the bottom of the cylinder, with smaller stones settling above them. The humus should float on the water.	Soil samples Measuring cylinder Scissors Glue	The amount of humus and soil particles will depend on the type of soil that has been used. Rich soils and compost will have a great deal of humus (dead plant and animal matter).
Sheet 9 (page 84) Types of soil		Soil samples Pestle and mortar Microscope/magnifying glass	The size of the soil particles will depend on the characteristics of the soil that is used. Sandy-based soil will tend to have large particles and clay soil will tend to have small particles.
Sheet 10 (page 85) Soils and drainage		Filter paper Funnels Measuring cylinders Timer Soil samples	The amount of water that drains through the soils will depend on the types of soils used and their existing water content.

Life Processes and Living Things

Living Things in their Environment

Activity	Answers	Resources needed	Teaching/safety notes
Sheet 11 (page 86) Food chains	• Wheat – mouse – owl. • Carrots – rabbit – fox. • Cabbage – caterpillar – sparrow.	Scissors Glue Reference materials	
Sheet 12 (page 87) More food chains	• Woodland: leaves – worm – mole. • Pond: beetle – perch – pike. • Field: wheat – sparrow – hawk.	Scissors Glue Reference materials	
Sheet 13 (pages 88 and 89) Food web	[Food web diagram: Owl at top; connected to Birds such as robins and Moles; further connected to Mammals such as mice, Insects, Caterpillars, Earthworms; bottom level: Acorns, Leaves and bark, Dead leaves]	Scissors Glue Reference materials	This food web is only one possible food web of the many that can occur in and around an oak tree.
Sheet 14 (page 90) Predators and prey	• Cat – mouse • Killer whale – seal • Bear – salmon • Lion – zebra	Scissors Glue Reference materials	
Sheet 15 (page 91) The importance of plants	**Food plants** Apples Strawberries Tomatoes Lemons Mushrooms Carrots Potatoes		
Sheet 16 (page 92) Plants and animals crossword	**Across** 3 shelter 5 fertilizer 6 sun 7 shade **Down** 1 disperse 2 water 4 plants 5 food	Reference materials	

Living Things in their Environment

Activity	Answers	Resources needed	Teaching/safety notes
Sheet 17 (page 93) Tooth decay 1	• Left, top to bottom: gum, nerve and blood vessels. • Right, top to bottom: enamel, dentine, pulp cavity.	Coloured pencils Reference materials	
Sheet 18 (page 94) Tooth decay 2	The correct order is: 4, 3, 1, 2.	Scissors Glue	

Life Processes and Living Things

Habitats

Living Things in their Environment – Sheet 1

You need: scissors and glue.

 The places where animals and plants live are called **habitats**. There are different types of habitat to suit different plants and animals.

Woodland habitat

Field habitat

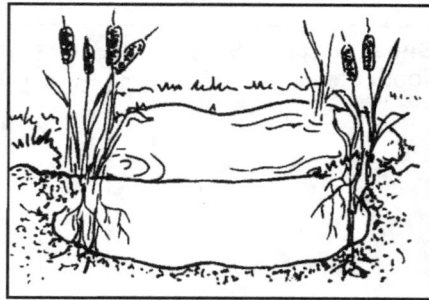

Pond habitat

- Cut out the habitats above and the animals below.
- Match the animals with their preferred habitat.

Squirrel	Hare	Goose	Mouse
Fox	Frog	Badger	Horse
Owl	Tadpole	Dairy cow	Perch

Life Processes and Living Things

© M Abraitis, A Deighan, B Gallagher, B Smith & M Toner
This page may be photocopied for use by the purchasing institution only.

Which habitats?

Living Things in their Environment – Sheet 2

 **Read** There are lots of different types of habitats. Plants and animals are found in the habitat which suits them best.

 A squirrel is most likely to live in woodland where it can climb trees and find acorns to eat.

 A frog is most likely to live in a wet, shady place like a pond or river.

 **Activity** Think of the animals below and how they live, then suggest the type of habitat they might prefer.

Animal	Preferred habitat
Otter	
Mouse	
Deer	
Greenfly	
Newt	
Badger	
Woodlouse	
Slug	
Robin	
Mole	
Grasshopper	
Worm	
Bee	

Some habitats

Pond or river

Field

Woodland

Under leaf or stone

Grass

Hedge

Flower bed

Tree

Living Things in their Environment – Sheet 3
Which habitat for woodlice?

> You need: a tray with some sand on the bottom of it, some dry leaves, some stones, some wet bark and some woodlice.

 **Activity** Create three different habitats by putting some dry leaves, some stones and some wet bark in a tray with sand at the bottom of it.

Which habitat do you think the woodlice will prefer?

Investigate which habitat the woodlice prefer by putting some into the middle of the tray and watching where they go.

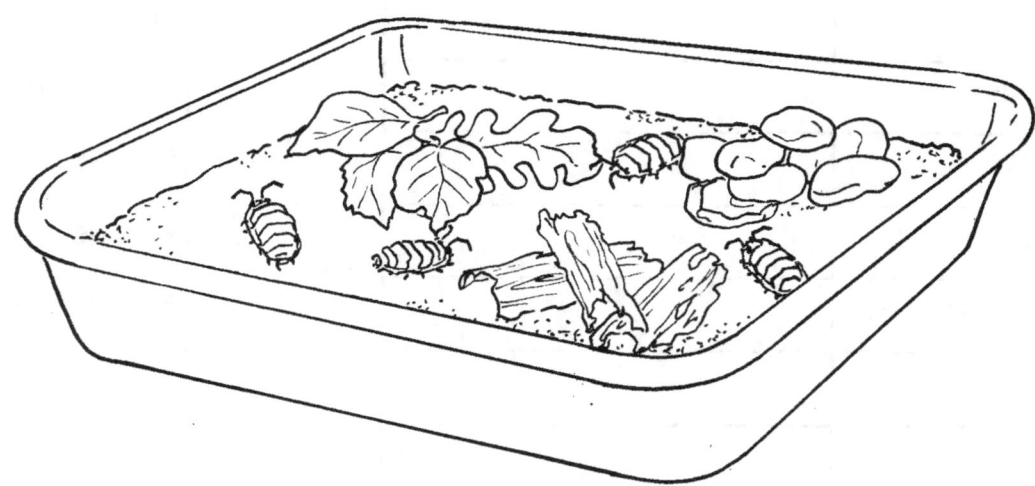

 **Answer** The woodlice prefer the _____ habitat.

Life Processes and Living Things

The wormery

Living Things in their Environment – Sheet 4

You need: a wormery, 6 worms, some leaves and cardboard.

 Animals live in habitats which suit them best. A wormery is a good way of studying worms in a habitat which is close to their natural one.

- Place some worms carefully onto the top of the soil and observe what they do.

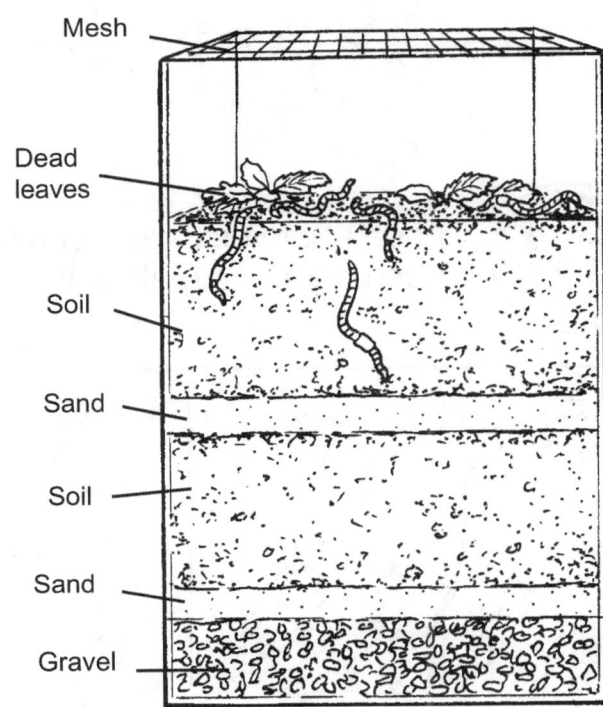

- Surround the sides of the wormery with thick cardboard and then leave it undisturbed for five days.

- While the worms are kept in darkness think about some of the things you could investigate about them.

Here are some ideas you could think about:

- Do the worms live on top of the soil or underneath it?
- What will happen to the leaves?
- What shape are the burrows?

 I will investigate...

Life Processes and Living Things

Animals in their habitat 1

Living Things in their Environment – Sheet 5

 **Read** Most living things require food, water, air, warmth and shelter from their habitat. Animals must be able to find food in their habitat or they will not be able to live in it.

Squirrels live in a woodland habitat where they can feed on nuts from the trees.

 **Activity** Look at the pictures below and describe how the habitats can supply each of the animals with their food.

1	2	3
4	5	6
7	8	9

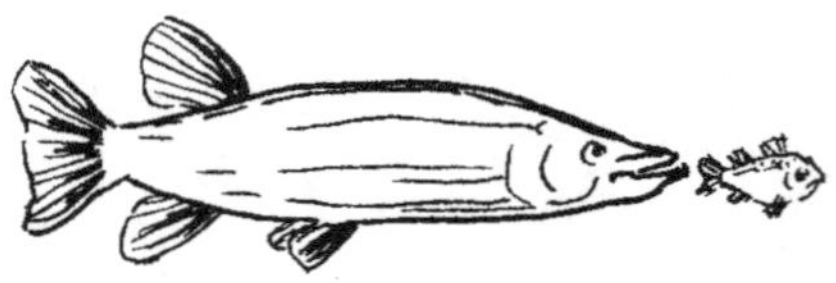

Life Processes and Living Things

Living Things in their Environment – Sheet 6
Animals in their habitat 2

Read

Animals and plants live in habitats which suit them best. Most hares live in grassy fields where they obtain food to survive.

- Strong leg muscles for running
- Large ears to listen for danger
- Sharp teeth for biting food

Activity

Label the features of the animals below which show how they are suited to their own habitat.

Helpful phrases

Webbed feet for swimming
Tail for balance
Waterproof feathers to keep warm

Strong teeth for biting hard
Sharp claws for gripping
Beak for gripping food

Adaptation

Living Things in their Environment – Sheet 7

Activity Polar bears and camels have features which make them suitable to live in their own very different environments.

Find out about some of these features and why they help the animal survive in its environment.

Animal	Environment	How is the animal suited to its environment?
Polar bear It has white fur, claws and sharp teeth. It is a good swimmer and is surrounded by thick blubber and fur. It has good eyesight and fur on the soles of its feet.	**The Arctic** It is very cold and has lots of slippery, white, floating ice. There are lots of seals, walruses and fish.	
Arabian camel It can survive for many days without water and has feet which splay out when walking. It has long, hairy eyelashes and nostrils which can be almost closed. It is covered with fur.	**The desert** Water can be hard to find and strong winds can blow sand and grit around. It is hot during the day and cold at night. The soft sand is hard to walk on.	

Life Processes and Living Things

Living Things in their Environment – Sheet 8

What is soil?

You need: soil samples, a measuring cylinder, scissors and glue.

 Soil is made from rocks (which are worn into tiny particles by the weather) and the remains of dead animals and plants (which is called humus).

 Investigate what the soil around your school is made from by trying out the following activity.

◆ Put a few teaspoons of soil into a measuring cylinder and then add about 30 cm³ of water.

◆ Shake the mixture and leave it to settle.

◆ After the mixture has settled, draw what you see on the empty cylinder below.

◆ Use the labels at the bottom of the page.

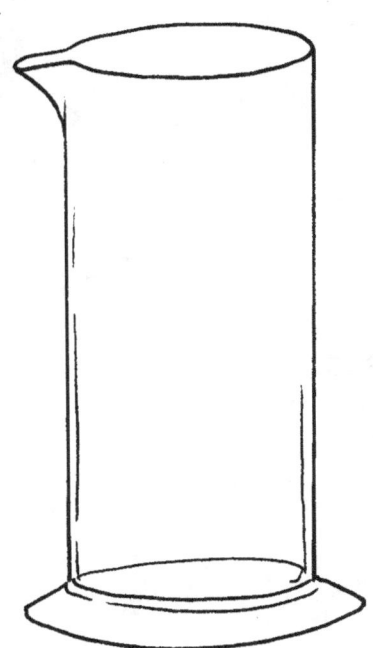

| Humus | Clear water | Large stones | Small stones |

Living Things in their Environment – Sheet 9

Types of soil

You need: some soil samples, a pestle and mortar and a microscope or a magnifying glass.

 **Read**

Plants are normally found growing in soil. Soil can be sorted into three types.

Sandy soil is made up of large particles which have large air spaces between them. Sandy soil does not hold much water.

Clay soil is made up of small particles which have a few air spaces between them. Clay soil becomes water-logged easily.

Loam soil is a mixture of clay and sandy soil. Loam soil holds just the right amount of water and is good for growing plants.

 **Activity**

◆ Collect some dry soil samples from around the school.

◆ Crush any lumps with a pestle and mortar and then examine some of the grains under the low-power microscope or magnifying glass.

 **Activity**

Draw what the grains look like.

What type of soil do you think it is?

Life Processes and Living Things

© M Abraitis, A Deighan, B Gallagher, B Smith & M Toner
This page may be photocopied for use by the purchasing institution only.

Soils and drainage

Living Things in their Environment – Sheet 10

> You need: filter paper and funnels, measuring cylinders, a timer and some soil samples.

- ◆ Investigate how long it takes water to drain through soils from different locations around your school.

- ◆ You will need to crush the soil samples and put equal amounts of soil into funnels with filter paper in them.

- ◆ Pour an equal amount of water through each of the samples and measure the amount of water which drains into the measuring cylinders.

	Sample 1	Sample 2	Sample 3
Amount of water collected			

What can you tell from your investigation?

© M Abraitis, A Deighan, B Gallagher, B Smith & M Toner

Life Processes and Living Things

Food chains

Living Things in their Environment – Sheet 11

You need: scissors and glue.

 All living things need to eat food to survive. A **food chain** shows how living things depend on each other for survival.

For example: **Lettuce leaf** → **Snail** → **Bird**

Most food chains start with a **green-leafed plant** and finish with an **animal**.

◆ Cut out and arrange the pictures below into three food chains.

◆ Glue the pictures onto a separate sheet of paper in their correct order, using arrows to link them.

Wheat	Owl	Field mouse
Rabbit	Carrots	Fox
Caterpillar	Sparrow	Cabbage

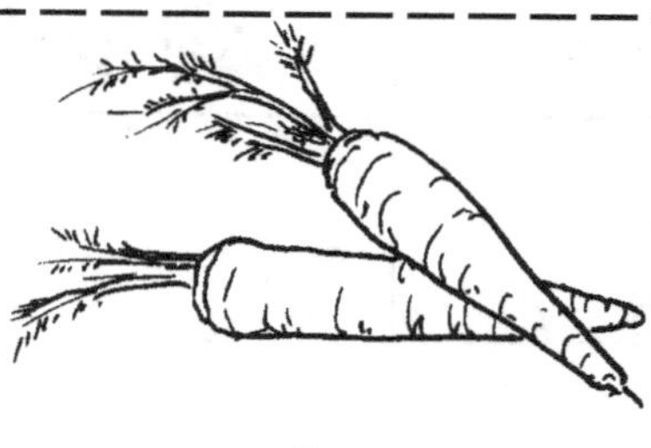

Life Processes and Living Things

Living Things in their Environment – Sheet 12

More food chains

You need: scissors and glue.

Activity

Cut out the pictures below and arrange them into food chains which could be found in a **woodland habitat**, a **pond habitat** and a **field habitat**.

Pike	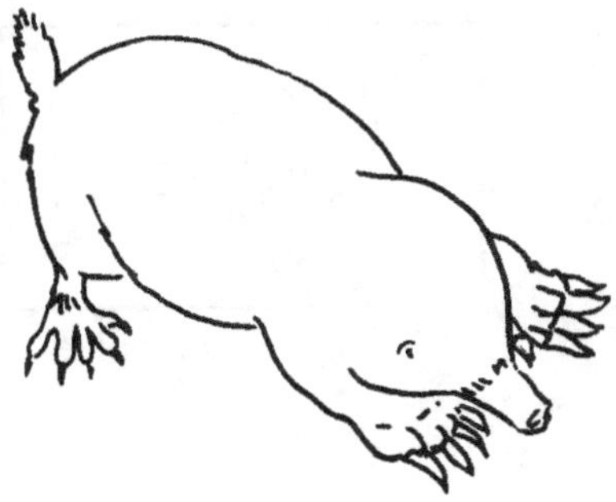 Mole	 Sparrow
 Sparrow hawk	 Water beetle	 Wheat
 Dead leaves	 Perch	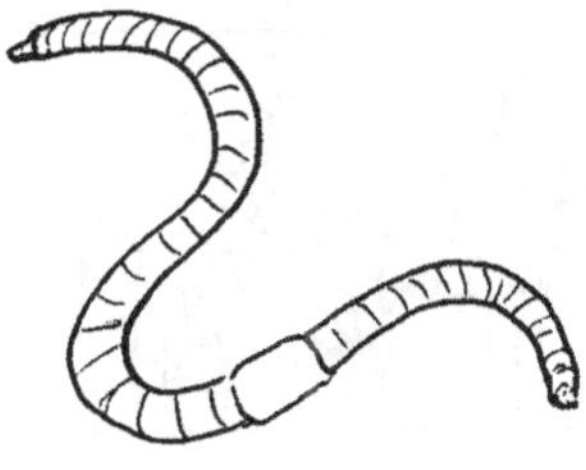 Worm

© M Abraitis, A Deighan, B Gallagher, B Smith & M Toner
This page may be photocopied for use by the purchasing institution only.

Life Processes and Living Things

Living Things in their Environment – Sheet 13
Food web

You need: scissors and glue.

Activity

More than one food chain in a habitat is called a **food web**.

Cut out the pictures below and stick them onto the oak tree as food chains to make up an oak tree food web.

Insects such as ants and beetles

Birds such as robins

An owl

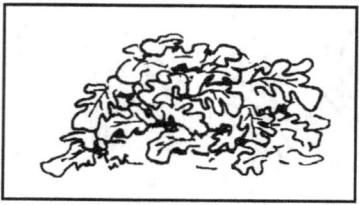

Dead leaves

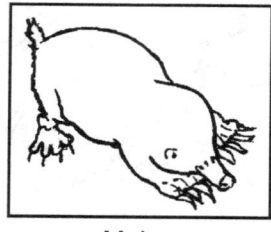

Moles

Earthworms

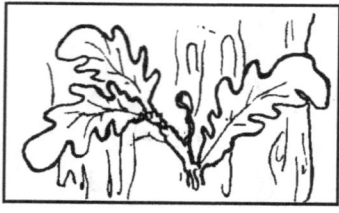

Leaves and bark

Acorns

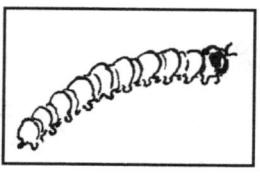

Caterpillars

Small mammals such as mice

Living Things in their Environment – Sheet 13 (Continued)

Food web

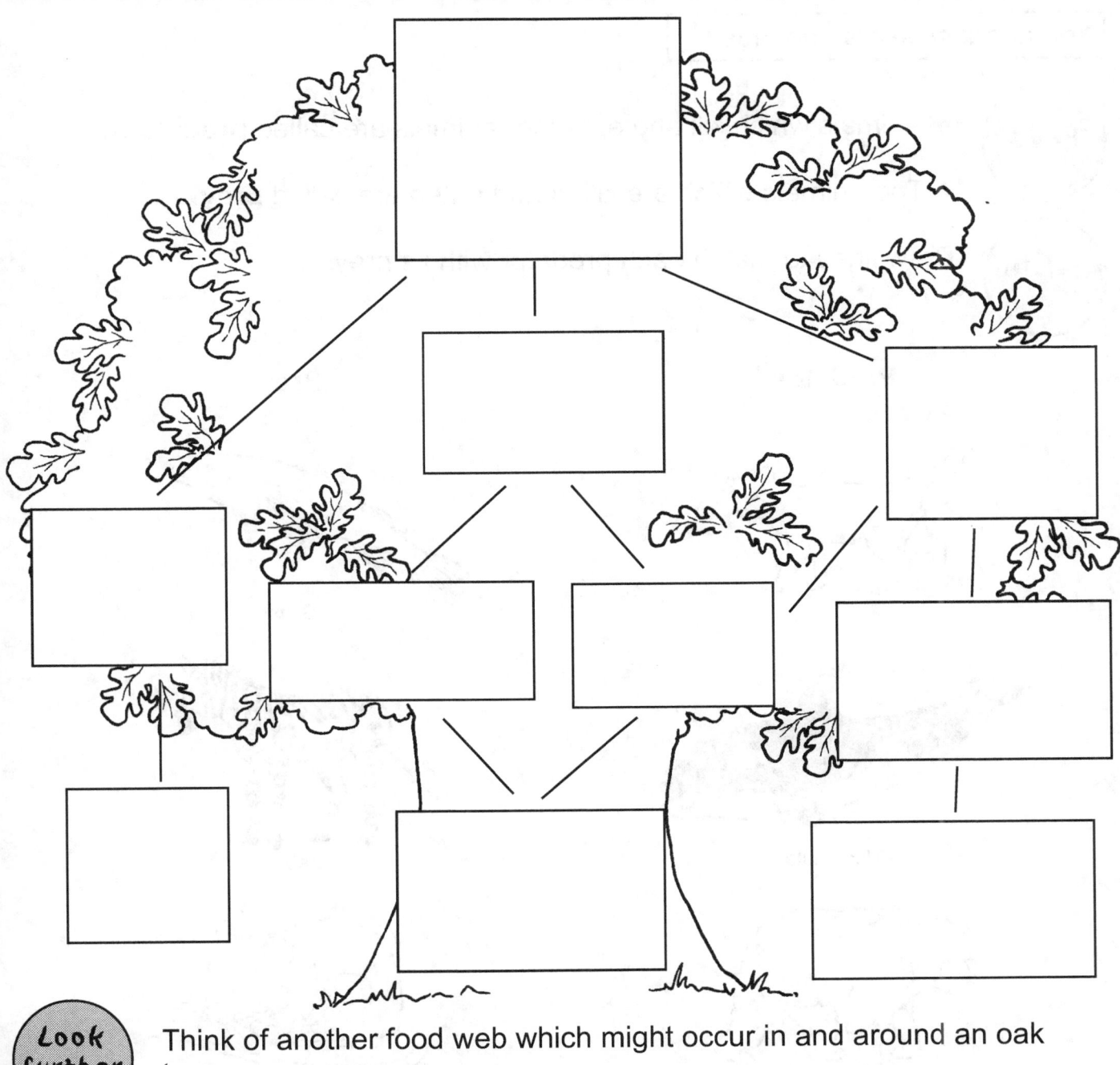

Look further — Think of another food web which might occur in and around an oak tree.

Predators and prey

Living Things in their Environment – Sheet 14

You need: scissors and glue.

- Animals which kill and eat other animals are called **predators**.
- The animals which are killed and eaten are called **prey**.

Draw lines to match each predator with its prey.

Predator   **Prey**

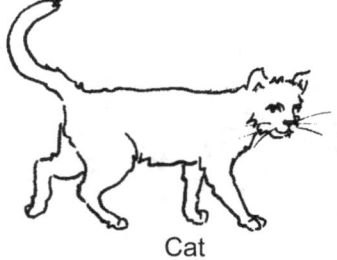
Cat

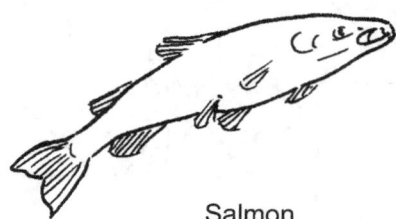
Salmon

Killer whale

Zebra

Bear

Mouse

Lion

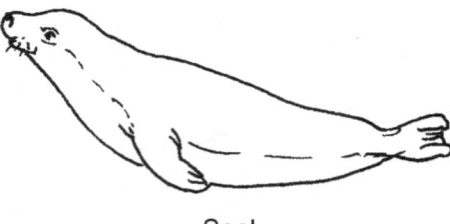
Seal

Living Things in their Environment – Sheet 15
The importance of plants

 **Read** — Although many plants are grown because they look nice and have a lovely fragrance, plants are also grown because they produce food which we can eat.

 **Activity** — Which of the plants below are grown as food?

Apples	Crocuses	Strawberries
Tomatoes	Heather	Lemons
Mushrooms	Carrots	Potatoes

Plants and animals crossword

Living Things in their Environment – Sheet 16

Across
3 Monkeys use trees for __ __ __ __ __ __ __ at night.
5 Animal droppings can be helpful to plants as __ __ __ __ __ __ __ __ __ __ __ .
6 The __ __ __ is the source of energy for green plants.
7 When lions are too hot they can use plants for __ __ __ __ __ .

Down
1 Many plants rely on insects and animals to __ __ __ __ __ __ __ __ __ their seeds.
2 All living things need __ __ __ __ __ to drink.
4 Many animals which are preyed upon feed on green __ __ __ __ __ __ .
5 Some animals rely on plants for __ __ __ __ to eat.

Life Processes and Living Things
© M Abraitis, A Deighan, B Gallagher, B Smith & M Toner
This page may be photocopied for use by the purchasing institution only.

Living Things in their Environment – Sheet 17

Tooth decay 1

You need: coloured pencils.

- Use books or a CD-ROM to help you label the tooth with the words below.
- Colour in the different parts of the tooth.

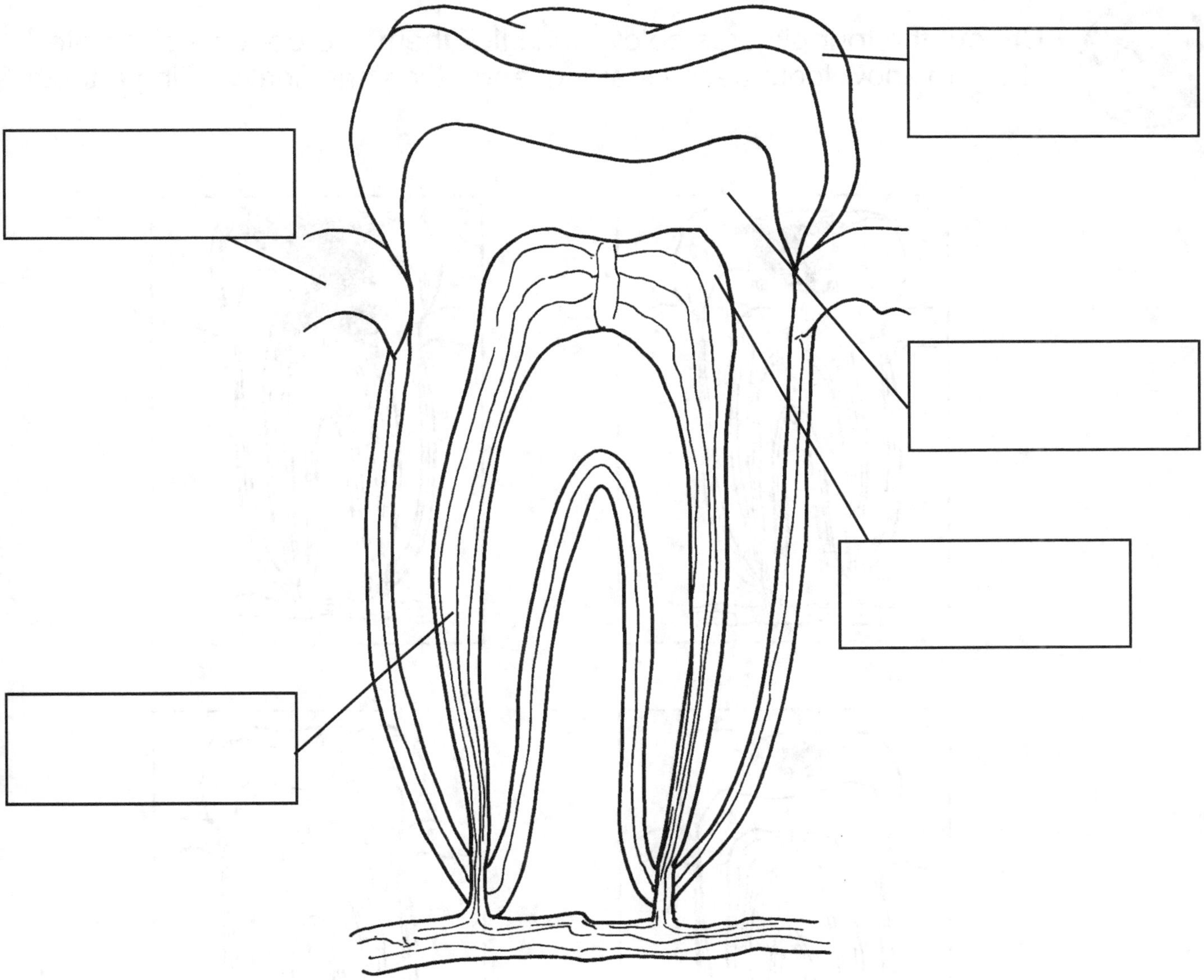

| Pulp cavity | Enamel | Gum |
| Nerve and blood vessels | Dentine | |

Living Things in their Environment – Sheet 18

Tooth decay 2

You need: scissors and glue.

After you have eaten, some sugar may remain on your teeth. Microbes eat this sugar and produce acid. If the microbes are not removed by brushing, then the acid will dissolve the enamel and then the dentine. If the dentine is missing, the nerve is exposed and causes toothache. The tooth then dies.

Cut out the four pictures below and stick them in order on a separate sheet to show tooth decay starting, spreading and finally killing a tooth.

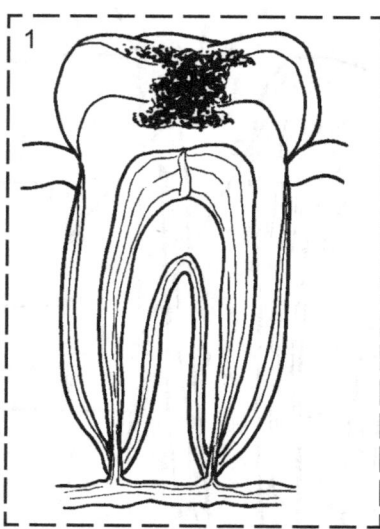
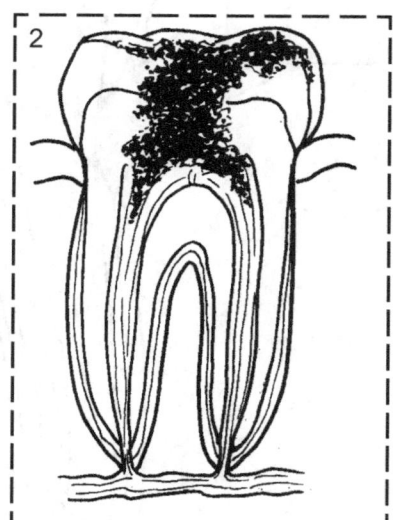
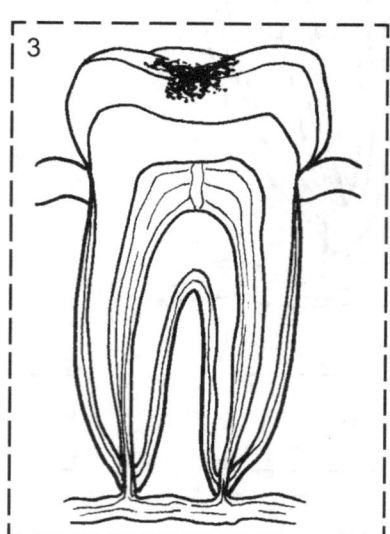
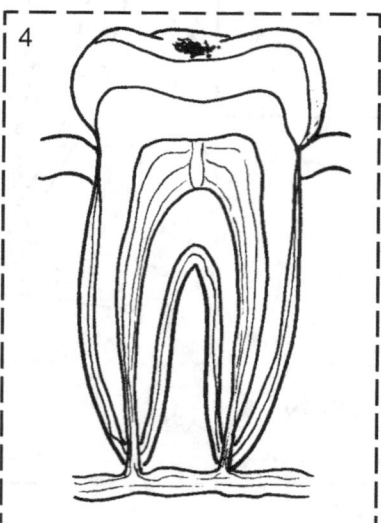

Life Processes and Living Things

© M Abraitis, A Deighan, B Gallagher, B Smith & M Toner
This page may be photocopied for use by the purchasing institution only.

Glossary

These definitions, while accurate, are by no means comprehensive. If you require a more exact definition, you should consult a good scientific dictionary.

amphibian
A vertebrate animal with smooth, moist skin.

arteries
Tubes which carry blood away from the heart to the rest of the body.

bacteria
Micro-organisms which looks like living cells, but they do not have a nucleus. Some bacteria are helpful, others are not.

bird
A vertebrate animal which has feathers and lays hard eggs.

canine
Sharp, pointed tooth for biting down on food.

carbon dioxide
An atmospheric gas used by plants during photosynthesis. Plants take in carbon dioxide from the atmosphere and give out oxygen back into the atmosphere.

cell wall
Gives the cell shape.

chlorophyll
The chemical which gives plants their green colour and which is essential for photosynthesis to occur.

chloroplast
This part of a green plant cell is where photosynthesis occurs.

consumer
An organism which eats another organism. Normally a primary consumer eats plants and a secondary consumer eats animals.

cytoplasm
This part of a green plant cell is where many chemical reactions take place.

dentine
Dentine is a living part of the tooth which is located under the hard enamel which protects the tooth.

ecosystem
The ecosystem is the way in which organisms are dependent on each other and the conditions of their environment for survival.

enamel
The white outer part of a tooth. It surrounds and protects the living part of the tooth.

environment
The conditions in a habitat, for example temperature and rainfall, make up the environment.

fertilization
When a male cell joins with a female cell.

fish
A vertebrate animal with wet scales.

fluoride
A chemical which some scientists believe hardens the enamel and helps protect the tooth from decay.

food chain
The way in which living organisms are dependent on each other as food.

food web
A food web is made up of two or more food chains.

fungi
Micro-organisms made from long thread-like structures called hypha.

habitat
A habitat is the place in which an organism lives, for example a woodland or a desert.

incisor
Tooth at the front of the mouth used for cutting food.

invertebrates
Animals without backbones, such as insects.

mammal
A vertebrate animal which has hair.

molar
Tooth at the back of the mouth used for chewing. The smaller molars are called premolars.

nerve
A living part of the tooth. The nerve gives the tooth sensation. A little too much sensation is toothache.

nucleus
The nucleus is the control centre of the cell. It contains DNA.

nutrients
Chemicals such as iron and calcium which are essential to life.

organism
Any living thing, cell, plant or animal can be described as an organism.

oxygen
An atmospheric gas. Plants give out oxygen during photosynthesis. Humans and other animals take in oxygen from the atmosphere when we breathe.

photosynthesis
The process by which plants turn light energy into chemical energy (food).

predator
Any animal which kills other animals for food.

prey
An animal killed or hunted by other animal(s) for food.

pulp
A living part of the tooth.

producer
Normally a green plant, such as lettuce, which produces food from the Sun's light.

protozoa
Micro-organisms which can cause disease.

reptile
A vertebrate animal with dry scales.

vacuole
This part of a green plant cell contains sap and other chemicals.

veins
Tubes which carry blood from the body to the heart.

vertebrates
Animals with a backbone, such as amphibians, birds, fish, mammals and reptiles.

virus
A type of micro-organism that cannot reproduce by itself. Viruses cause many diseases such as flu, chicken pox, cold sores and HIV.

www.ingramcontent.com/pod-product-compliance
Lightning Source LLC
Chambersburg PA
CBHW080054200426
43197CB00053B/2722